# The Wired Neighborhood

# The

# Wired  Stephen Doheny-Farina

# Neighborhood

Yale University Press

New Haven and

London

Published with assistance from the foundation established in memory of William
McKean Brown.

Designed by Nancy Ovedovitz and set in Adobe Garamond type by Tseng Information
Systems, Inc. Printed in the United States of America.

Library of Congress Cataloging-in-Publication Data

Doheny-Farina, Stephen.

The wired neighborhood / Stephen Doheny-Farina.

p.      cm.

Includes bibliographical references and index.

ISBN 0-300-06765-8 (hc : alk. paper)

1. Information superhighway—Social aspects.    2. Internet (Computer network)—
Social aspects.    3. Virtual reality—Social aspects.    I. Title.

HE7568.D64    1996

303.48'34—dc20            96-12241

                         CIP

A catalog record for this book is available from the British Library.

Printed on recycled paper.

The paper in this book meets the guidelines for permanence and durability of the Committee
on Production Guidelines for Book Longevity of the Council on Library Resources.

10   9   8   7   6   5   4   3   2   1

# Contents

## The Immersions

## A Preface

*Diving.* A man in a black wet suit slid himself across large rocks toward the ocean. I had just finished suiting up, and I glanced down at him to gauge how slippery the rocks would be in the descent to the waterline. A few minutes earlier I had been watching as the man sat on a nearby rock while his assistant helped him put on his tank, buoyancy compensator, and weight belt. All the while the man's wheelchair sat motionless and empty on the edge of the parking lot. We were both about to go diving in the cold waters by Nubble Lighthouse on the Maine coast, but the entry to the dive was difficult. As I started to ease my way down among the mass of wet boulders and seaweed, I realized that our relative mobility was beginning to equalize. No one wearing flippers, tank, and a weight belt could walk to the water across this terrain. He pulled himself with his arms; I sat down and slid from rock to rock. By the time we were both immersed in the ocean, technology and nature had pretty much leveled our ability to explore.

*Soaring.* I stood with my wife at the back end of Pit Road minutes before a race sponsored by the National Association for Stock Car Auto Racing at the Charlotte Motor Speedway. Ahead of us, lined up two by two, sat the cars, idling, shaking, and roaring as pit crew members in brightly colored uniforms and ear protection dashed about, making last-minute adjustments. We could barely hear the pronouncements over the booming public address system, but we knew they had something to do with Memorial Day and remembrances for our veterans. Thousands of black and white balloons were about to be released to honor those missing in Vietnam. On the infield grass a band and chorus were performing patriotic songs, which added more clutter to the deafening noise that had engulfed us.

We, too, were part of a crew, but not for one of the cars. We were the ground crew for a vehicle that was still a couple of miles away. And we knew something the hundred thousand fans in attendance did not know. Within minutes, unannounced, a glider—a sleek soarplane—would suddenly and silently swoop down along the straightaway that runs the length of the massive grandstands. It would come as low as it safely could, then would rise up over the far stands, circle back around, and land on the track across the starting line. The pilot, Terry, was a colleague of mine, a former soarplane racer, and a Vietnam veteran. His task was to deliver the green flag for the start of the race. The ground crew's job was to retrieve Terry and the plane and get it off the track and behind Pit Road as fast as possible.

The plan worked flawlessly. I was told later by people in the stands that the glider caught everyone's attention and that most people were riveted when they saw someone (me) pushing a wheelchair as fast as he could across the infield grass while the pilot opened the cockpit, waved the green flag, and pulled himself out of the plane, revealing his disability. As the public address system blared Terry's story—"lost his legs in a mortar attack in Vietnam"—we were all immersed in the noise, the technology, and a massive infusion of ideology.

*Dreaming.* Every Tuesday and Thursday afternoon from two o'clock until half past four, quadriplegic and paraplegic patients at the Bronx Veterans Affairs Medical Center roll their wheelchairs up to computer terminals, put on head-mounted displays and data gloves, and immerse themselves in virtual worlds, as N. R. Kleinfeld describes in a *New York Times* article. Once there, they have the sensation of moving and acting in a three-dimensional space where they are no longer disabled. The program is the result of work done by William Meredith, a recording engineer in New York City who for a decade has been helping disabled veterans at the center on a volunteer basis. After years of trying to raise the spirits of paraplegics and quadriplegics, Meredith wondered what virtual reality (VR) could do for them. "These visions ran through my mind. These people could fly, which they can't. They could walk, which they can't. They could play sports, which they can't."

One paraplegic veteran commented, "You know why a lot of veterans are in and out of hospitals? Stress. If they want to have any dreams, they have to get them from a bottle. Here, you can have dreams without the bottle. All I can do is look here and see a lot of potential. An angel with a lot of wings." And Meredith anticipates expanding the reach of these virtual worlds: "Ultimately, I want to have interactive sports. I'd like to link up several hospitals and have leagues and everything. They'll play baseball, football, whatever they want. They'll be able to feel every hit."

All three of these stories show individuals using technology to liberate themselves and others. As I have reconstructed them, they are hopeful, offering us glimpses of ambition, tenacity, and goodwill. But the temptation is to mistake the social and economic forces enabling these liberations for the qualities of the technology. That is, it is tempting to say that scuba diving, gliding, and VR technologies are *inherently* liberating.

These days we are all immersed in waves of right thinking about technologies like VR, the Internet, and broadband, global, multimedia communication networks. Techno-experts tell us we are living in revolutionary times as we witness the shift from the industrial society to the knowledge society. The cheerleaders of the information age are continually inviting us to be ecstatic about this shift. Conversely, others are trying to persuade us to reject the new technologies. They are the voices of the "cyberspace backlash" or the "neo-Luddite movement," which urge us to turn off our electronic screens and maybe even smash our computers to bits—not the bits that may be stored on our hard drives, but the kind that scatter across a table and onto the floor, the kind you can't put back together again.

In *The Whale and the Reactor,* Langdon Winner argues that technologies are not neutral but embody specific forms of power and authority. In the 1990s, debates about the nature of the power and authority of the new communication technologies have been polarized by the discourse of the techno-utopians and the neo-Luddites. The issues, of course, are more complicated than that, as Gary Chapman notes in his article "Making Sense Out of Nonsense: Rescuing Reality from Virtual Reality." Sophisticated computer-based technologies, Chapman observes, are quite difficult to characterize. Although they are not neutral vessels, their natures are evolving, complex, and often indeterminate.

Accordingly, this book falls in neither the techno-utopian nor the neo-Luddite camp. My view of our emerging electronic communication technologies is highly critical, but the target of that criticism is not some inherently negative quality possessed by those technologies. The real target is a vision that fuels much of the work with the technologies, a vision that can be summed up by a quotation attributed to Louis Rossetto, the publisher of *Wired* magazine, in a 1995 *New York Times Magazine* article by Paul Keegan: "We're talking about the beginnings of exo-brains," he confides. "Brain appliances. And exo-nervous systems, things that connect us up beyond—literally, physically—beyond our bodies, and we will discover that when enough of us get together this way, we will have created a new life form. It's evolutionary; it's what the human mind was destined to do."

An even grander but similar sentiment opens an electronic document known as the "Magna Carta for the Knowledge Age," written by four high-tech luminaries: Esther Dyson, George Gilder, George Keyworth, and Alvin Toffler. In the preamble, the authors make the following pronouncement: "The central event of the 20th century is the overthrow of matter. In technology, economics, and the politics of nations, wealth—in the form of physical resources—has been losing value and significance. The powers of mind are everywhere ascendant over the brute force of things." *Deny the body,* proclaim the leaders of this revolution. Join us and we will all achieve quadriplegic nirvana. Furthermore, if we are going to deny physical matter, we must deny geophysical place and time. Again, the "Magna Carta": "It is clear . . . that cyberspace will play an important role in knitting together the diverse communities of tomorrow, facilitating the creation of 'electronic neighborhoods' bound together not by geography but by shared interests."

In response to such electro-evolutionary visions, I argue that we do not need electronic neighborhoods; we need geophysical neighborhoods, in all their integrity. The revolution that must be joined is not one that removes us from place but one that somehow reintegrates the elements of our dissolving placed communities. But that is a daunting and often unglamorous task, which may not fall within the parameters of the electronic revolution. "The wish to leave the body, time, and place behind in search of electronic emulation," say James Brook and Iain Boal in their preface to *Resisting the Virtual Life,* "does not accidentally intensify at a time when the space and time of everyday life has become so uncertain, unpleasant, and dangerous for so many—even if it is the people best insulated from risk who show the greatest fear." The electrosphere becomes a refuge for those who wish to avoid people who are different from themselves. Jon Katz, for example, in a review of Kirkpatrick Sale's *Rebels Against the Future: The Luddites and Their War on the Industrial Revolution,* states that Sale, in his neo-Luddism, ignores the constructive nature of the net while engaging in an uninformed argument about the ways that computers separate people from one another.

This is perhaps the most common of the conventional media bugaboos about digital culture: the chilling What-Kind-of-Country-Will-It-Be-When-

Everybody-Is-Staring-into-a-Screen-All-Day-Instead-of-Touching-and-Meeting-One-Another fear.

Probably the first thing anybody notices when they go online, however, is the community-building taking place all through cyberspace. Old people talk to old people, lonely gay teens find other lonely gay teens, unpublished poets trade poems with unpublished poets, physicians swap case histories with physicians. . . .

On computer conferencing systems and bulletin boards, thousands of virtual communities have sprung up across the world, enabling disconnected people to communicate with one another. Perhaps more than any other element, this is the truest miracle of digital information. Sale doesn't even mention it in *Rebels Against the Future*.

Nor does he notice one of the more ironic connections between the Industrial Revolution and the digital one: if it first pulled people from their families, computer technology is returning them in droves. More Americans than ever before now work from home, mostly because of digital fax-modem capability.

Katz thus asks, How can someone damn computers when they deliver all these life-affirming, life-enhancing things? In this book I don't damn computers, but I examine the validity of each of those supposedly life-affirming and life-enhancing qualities that Katz mentions in his defense of the technologies.

Yet, even while I devote most of the book to this attack on the techno-utopians, I do not lay claim to a neo-Luddite position that characterizes technology as evil. Instead, I attempt to stake out a portion of the less elegant and much messier middle ground. On one hand, I argue that powerful cultural trends are seducing us toward technological immersion and away from our placed lives. I thus have a negative view of much of what I see on the net. On the other hand, I recommend not shunning the net but steering it.

At the end of *The Future Does Not Compute*, Stephen Talbott repeatedly asks whether "technology has us in a stranglehold." He pushes us to consider whether there is "any future free of the machine's increasingly universal determinations." In this book I, too, am concerned about transcending the new communication technologies, but I am just as concerned that we devise and maintain ethical uses of them—such uses as those found within the civic networking movement.

Civic networking describes limited, focused, carefully applied efforts that attempt not to move us into cyberspace but to use communication technologies to help reintegrate people within their placed communities. In that sense it pushes against the continual virtualization of everyday life. But these efforts at community development can also suffer from their own forms of utopian hype. Accordingly, in this book I attempt to paint a balanced, sober picture of the daunting task of countering the destructive, globalizing seductions of cyberspace. As such, then, this book is another drop in the trickle of doubt about the communication revolution we are living through—a drop immersed, I'm afraid, in torrents of hype and hope about the glorious overthrow of matter.

In this book I use several terms in unusual ways and cite some unpublished materials, as described below.

*Virtual Reality and VR.* As I use these terms, "virtual reality" can be induced to varying degrees through any communication technology— print, film, video, text-only computer networks, and so on. But VR is induced only by immersive vehicles of image and sound—the head-mounted display–data glove technologies and all their variations. As we move from old technologies like print to new technologies like VR, we increase immersiveness and interactivity. I do not cede immersion and interactivity to VR only, however; that is, I believe that all communication technologies enable these qualities, to different degrees.

*The Net.* Most often I use this term to refer to the vast mix of informal and formal electronic communication networks that encompass a range of technologies like fax machines, telephones, broadcast and cable mass media, satellite and cellular communications, computer networks, and so on. At times, however, I also use the term to refer not to that entire jumble of connections but instead to one component, the Internet (itself

a vast jumble). Sometimes I distinguish one definition from the other;
sometimes it doesn't matter.

*Names, Pseudonyms, and Anonymity.* I have substituted pseudonyms for
the names of most of the people about whom I write, including all the
character names I have identified through my experiences in text-based,
networked virtual realities like Diversity University MOO and MediaMOO.
All statements attributed to online characters in MediaMOO are fictional
—except for one—and used solely for illustrating an aspect of the communication technology. The only direct quotation I have included was
uttered by an anonymous guest with whom I communicated while I, too,
was connecting as an anonymous guest during my first MOO experience.

*Internet Sources.* Many of the sources cited in this book are available
via the Internet. Although I have tried to provide accurate electronic
addresses for all these citations, it is possible, given the nature of the
Internet, that some archives may have changed or may no longer exist or
that some addresses may have changed or may no longer exist. To help
readers conduct further research, I have included (with permission) a list
of Internet-based organizations whose work is relevant to the issues raised
in this book.

My sincerest appreciation goes to my colleagues Bill Vitek, whose ideas
about community and place were the catalyst for this project, and Greg
Clark, whose critical readings helped me shape this book. I should also
like to thank for their help and advice the following people: Jean Thom-
son Black and Jenya Weinreb of Yale University Press, Mike Clark, Mark
Curran, Lauren-Glenn Davitian, John December, Pete Deuel, Glenn
Ellis, Sonia Ellis, Geoff Farina, Susan Grasso, Dennis Horn, Bill Karis,
Susan Ross, and Stuart Selber. Further thanks to my colleagues Laura
Gurak and Jim Zappen and the rest of the participants in an online seminar, "Rhetoric, Community, and Cyberspace," whose counter-arguments
to my positions always kept me on my toes: Chris Barrett, Mary Barto-
senski, Chris Boese, Allan Heaps, Kevin Hunt, Patricia Jackson, Karla

Kitalong, Don Langham, John Logie, Lisa Mason, Will Powley, Geoffrey Sauer, Marilyn Urion, and Kim Wharton.

The research that informs Part IV of this book was supported in part by a 1994–95 Canadian Studies Research Grant, Canadian Embassy, Washington, D.C. Several brief segments of this book appeared in an earlier form in the "Last Link" column of the online *Computer-Mediated Communication Magazine* (http://sunsite.unc.edu/cmc/mag).

Part One

**Lost in the**

**Solitude of My**

**Virtual Heart**

Not a wired culture, but a virtual culture that is wired shut: compulsively fixated on digital technology as a source of salvation from the reality of a lonely culture and radical social disconnection from everyday life, and determined to exclude from public debate any perspective that is not a cheerleader for the coming-to-be of the fully realized technological society.

—*Arthur Kroker and Michael W. Weinstein,* Data Trash

# 1 Real Cold, Simulated Heat: Virtual Reality at the Roxy

The storm door slammed, rattling its window and echoing across the frozen yard. As soon as the door left my hand, we knew how cold it had gotten that evening. The gauge was the automatic door closer — the hydraulic tube and shaft designed to enable aluminum storm doors to glide to a close. We had long since discovered that this closer ceased to operate whenever the temperature dropped below zero degrees Fahrenheit. We had learned how to read the information provided by this technology because it had been sending us the same simple message for weeks. Night after night the temperature dropped into negative digits. It was January 1994, and our north country, the northernmost region of New York State, was in the middle of a record-breaking stretch of frigid weather. On the coldest night the temperature fell to forty below; some days it never rose above minus ten.

I let the door slam that night because my wife and I were in a hurry, as usual. It was the first night of the Cinema 10 film series at the Roxy, our town's lone theater, and we expected a crowd. Ten Monday nights in the fall and another ten in the winter gave the locals a chance to see movies that would never come to town otherwise. Tonight it was Kenneth Branagh's production of *Much Ado about Nothing*.

In this cold, the snow crunches and squeaks beneath your boots, the slightest breeze threatens to freeze exposed skin, and the car heater takes forever to get warm. But none of it mattered this night. We were bundled up enough to withstand the elements for the five-minute ride into town. When we arrived, a line was forming down the street. Everyone was battling the cold. Some huddled with their companions; some stood hunched, arms crossed, trying to hide mitten-clad hands beneath each arm; some bounced from foot to foot. Our collective breaths hovered visibly above us. Finally the doors opened, and we all shuffled inside.

My town is small, and I saw many familiar faces: college students and faculty, some local joggers—people I knew from Frozen Foote, a series of winter road races in a neighboring town—and a few members of the local bicycle club and the local chapter of the Adirondack Mountain Club. In retrospect, I wonder if everyone there was seeking psychic help in preparation for the long dark expanse of the real winter, the winter that stretches on and on after the holidays, the winter that isn't over when you proclaim it over (as I did that year by deciding to do no more cross-country skiing after a warm day in early April even though weeks of good skiing were left in the woods; I was trying to will the end of winter after being seduced by the false spring of the north country).

There was escape for all of us in the theater that night. The film reveled in warmth. It showed sun and sweat and golden, tanned limbs. It bathed us in oranges and yellows and reds. We waded through the lush green of the ripe Italian vineyards, and the sunshine washed over and through everyone. I can still see myself sitting there, dazed by the virtual heat. And the opposition between that heat and the real cold of the night paralleled a number of the tensions in the film: the camaraderie within the group versus the alienation of the individual; the celebrations of a community in the wake of its victory over an external necessity; a simple, unassuming, blind love versus a skeptical, complex, but ardent love; trust and faith versus cynicism and lies.

When it ended and the houselights came up, I recall that others looked the way I felt: smiling, sunstruck, and warm as we all began to fumble with scarves and hats and gloves. On my way out, I saw one of the organizers of the series still sitting as she pulled on her coat. I told her I had just spent a couple of hours in Italy. She nodded "me too."

The meaning of this intense experience of virtual reality must be assessed in terms of that night in that place among those people. For me, a seemingly individualizing experience—watching a screen in a darkened theater—became a small communal act because it was situated in and spoke to a common necessity: the need to get through this long, dark, unremitting winter.

(I am troubled by my recollections. The weather was real, I know, but was the community? Did I experience community only because I saw a depthless reflection of it on the screen? Am I constructing a community in the act of describing it?)

This was an unusual cinematic experience, not because the film was unusual—it wasn't—but because it was so connected to an extraordinary communal constraint. This happens rarely. I can remember only one other such experience: in the last week of March 1979 I was teaching school about fifteen miles from the Three Mile Island nuclear power plant when the plant suffered its infamous partial core meltdown. The incident happened midweek and became known to the local population through a small news item of little note. On Friday morning, however, it suddenly became a crisis. The governor announced a state of emergency. We heard that Harrisburg Hospital was evacuating patients. We were told that a radioactive bubble of pressure was developing within the reactor containment building. No one was sure whether the structure would hold. Within twenty-four hours, tens of thousands of area residents voluntarily evacuated. Schools, banks, businesses, restaurants, and most bars closed.

I had grown up in the area, and while my immediate and extended family headed out of town, a buddy of mine, Bill Weiss, and I decided to stay. (We loaded his van with food, water, and sleeping bags nonetheless. "Just in case it blows," we said.) Coincidence of coincidences: the just-released movie *The China Syndrome* came to the area that weekend. One local theater remained open to show it, and Bill and I went that first night. It may have been my earliest postmodern experience. Although the theater was packed (NBC News was there, covering the opening), every other business nearby was dark, and the streets were empty except for the traffic of the couple hundred moviegoers. We seemed to be wandering through a ghost community, accompanied only by those of us compelled to seek our reflections on the screen.[1]

Indeed, throughout the movie there were murmurs, bouts of nervous laughter, and moments of vocal recognition among us. At one point, when we finally came to understand exactly what had happened to the nuclear plant (run by a sweating, nervous Jack Lemmon), Bill leaned over to me and whispered, "That's not as serious as what happened here!" At another point, we all roared when a character representing a nuclear power expert stated that a meltdown could wipe out an area "the size of Pennsylvania." Yes, there we were, cheering our doom in unison, bound by an external necessity so vast and so terrible that we could do little but laugh at it. After all, we were coming to realize that we might need to evacuate the area permanently at any moment. So, regardless of the technical hindsight that validated the strength of the containment system, that night we faced the possibility that nuclear technology would destroy, forever, our community.

The two cinematic experiences I've described represent ephemeral moments of social bonding—borne out, yes, through media images; nonetheless, these were communal experiences made rich by place, the particular physical, geographic locations in which they occurred. Yet these moments are wholly insignificant compared with the powerful forces of electronic communication and electronic media that both individuate and globalize—forces that work to isolate individuals by exalting individuality, while making those individuals dependent on mass markets and globalized communication networks.

This is a subtle but devastating finesse: we become the globalized individuals, focusing on our individuality while becoming ever more reliant on large-scale markets and technologies. Further, in response to our isolation, we attempt to buoy our fragmented selves through artificial means of commitment and community. This is the result of the virtualization of everyday life and the concurrent demise of geographically bound, physical communities.

In *Habits of the Heart* Robert Bellah and his co-authors discuss Alexis de Tocqueville's examination of America in the nineteenth century.[2] According to Tocqueville, Americans place the rights of the individual above the rights of the collective. Yet the collective, the democracy, can thrive as long as individuals subscribe to a set of social mores, the "habits of the heart" that ensure the survival of society. Tocqueville nevertheless cautions about this powerful individuating tendency in the American character. He warns that this democracy enables a dangerous inward turn through which citizens forget their history and ignore their communities. Such individualists

> owe no man anything and hardly expect anything from anybody. They form the habit of thinking of themselves in isolation and imagine that their whole destiny is in their own hands.
>
>    Thus, not only does democracy make men forget their ancestors, but also clouds their view of their descendants and isolates them from their contemporaries. Each man is forever thrown back on himself alone, and there is danger that he may be shut up in the *solitude of his own heart*.[3]

It is no longer American democracy that isolates the individual; it is the simulacrum of democracy, the electronic democracy, the virtual culture, the society of the net that isolates individuals while seducing them with mere appearances of communication and collectivity. Once we begin to divorce ourselves from geographic place and start investing ourselves in virtual geographies, we further the dissolution of our physical communities.

Clearly, much in our society convinces us to do just that. The message is current and pervasive. John Markoff, for example, describes the phenomenon of climbing a peak in the Adirondacks while talking business on a cellular phone. He tries to make us feel good about how electronic communication enables us to be separated from the constraints of physical location:

> In the 1950's the sociologist David Riesman wrote in his book "The Lonely Crowd" that the death of community meant that one could be surrounded by people yet still be profoundly alone and isolated. Wireless communications technologies are turning his original vision inside out. With cellular phones and wireless E-mail, one can be physically alone yet still in the midst of a clamoring invisible crowd.
>
> "The community has triumphed over the individual," said Michael Schrage, a research associate at the Sloan School of Management at the Massachusetts Institute of Technology and a technology columnist for The Los Angeles Times.[4]

But, in reality, electronic communication pushes in exactly the opposite direction—toward the shadow we call virtual community. In immersing ourselves in the electronic net, we are ignoring our real, dying communities. The cinematic events I described illustrate how my geographic place impinged on and gave meaning to virtual experiences. But more than I may ever realize, the virtual mediates my understanding of my local, physical experience.

I am drenched in virtual worlds. One day I walked into a small open-air courtyard within a university building complex. As I moved through the archway into the courtyard, I could see that I was alone; no one was sitting on the benches among the trees and ornamental shrubs; no one was meandering through the sculpture garden. At the instant I stepped into the space, I heard a noise and looked up. A flock of birds flew over the courtyard, covering entirely its skyward opening. They were silhouetted against the sky and framed by four walls, and I heard the beating of their wings and then they were gone. My god, I thought, that was like an advertisement for Obsession perfume. Clearly, I see the world through a veil of images constructed through artificial media. This veil permeates all of my perception.[5]

It is fall as I write this, and the birds in the north country are mustering like troops, soon to fly south. Recently, as I walked along my road, I passed beneath a tall maple, about eighty feet high. Hundreds of birds perched, chirping, among its yellowing leaves. I don't often hear so many birds vocalizing at once at such close range. But I am so embedded in virtual worlds that it reminded me of the sounds I've heard at the Nature Company in the Carousel Centre Mall in Syracuse. On the few occasions when I've been to that mall I have visited the store because it markets its products through the use of sounds, images, and space to create a virtual atmosphere of nature: a rushing stream, its waters cascading down a rocky bed only to disappear into an unseen and silent pumping system behind a wall; CDs playing the cries of loons, wolves, or whales, or the gusting howls of the wind. On display are large posters of plants, animals, and mountains so clear and grand that they would appear real if framed within faux windows. A few strolls through this simulated natural environment have apparently informed the ways I experience the world. And this is merely a store at the mall. Imagine the impact of an immersion like the one envisioned by Steve Pruitt and Tom Barrett in their anticipation of a future workplace bound by virtual reality.[6]

Pruitt and Barrett derive a fictional account of work in Corporate Virtual Workspaces (cvws) in order to better understand the capabilities of networked virtual realities. In their story, a man named Austin enters a Personal Virtual Workspace (pvw) in a room of his California home, where he puts on computerized clothing and connects via fiber-optic network to the cvw "where" he works. The room and clothing interfaces enable him to see, hear, and feel everything in the virtual work environment. He is one of many employees connecting to the CVW from their "home reality engines." But the CVW is not exactly like an office building. Inside the CVW, Austin can design his own work space and tools to meet the needs of his current project and to suit his tastes.

At one point in his day, Austin must meet with a man named Johann who lives in Bonn, Germany, and who also connects to the CVW from his home. Austin virtually walks down the hallway to Johann's office, sees the door ajar, knocks, and walks in.

> The office is quite dramatic. Johann is obviously an avid mountain climber. One entire wall of the eight-sided office sports a lifelike panorama of El Capitan from Austin's home state of California. At first glance, it appears that the wall is just a still photograph of the majestic climbing magnet, but as Austin investigates it more closely, he notices that the leaves on the trees in the foreground are fluttering as if the wind were blowing. Upon still closer inspection, he notices brilliantly colored objects about halfway up the side of El Capitan. By invoking the zoom feature with a gesture toward a telescope icon in the lower-right corner of the wall, Austin brings the scene progressively closer. With another gesture toward a stop sign icon, he stabilizes the picture at about 50 yards from the band of hearty climbers that are making their ascent.
>
> The event is visually exciting and so is the audio content. By gesturing again at the wall's icon controls, the sounds of the scene are now audible via the directional sound receptors in Austin's computerized clothing. He hears the calls of "belay on" and "on belay" as the second climber clad in a burnt orange rugby shirt, dark brown knickers, and sky blue climbing shoes makes his way toward the lead climber positioned some 25 feet above at the next pitch. The wind whistles behind Austin as he hears the calls of a distant songbird.
>
> Suddenly, Austin's serenity is partially broken, as Johann reenters the CVW and appears between him and the El Capitan experience.[7]

After reading this passage, I'm left with a few questions: In his spare time is Johann a climber? Or is Johann a virtual climber? And if he is a virtual climber, does he actually don the computer suit and simulate the visual, aural, and tactile elements of the climb by scaling some structure that enables the simulated trek up El Capitan? Or does he merely watch the projection of himself climbing El Capitan while his suit and pvw provide stimulations, making him feel as if he is moving his body while he is actually sitting on the couch? Can Johann climb virtual El Capitan by himself with some sort of intelligent agent—a simulated partner—as his companion? More important, does Johann still know the difference between his virtual El Capitan and the rock geographically located in Yosemite? Most important, does it even matter? Does anyone need or want to travel to Yosemite to do the climb, and how is that climbing experience different after doing the virtual climb? Can Johann see the natural world anymore?

Because Johann is a fictional character, I'm going to supply some plausible answers: Johann is a virtual climber because all the real climbs are impossible now that poor air quality makes real climbing too dangerous. Johann's father did real climbs until his death at an early age from cancer. Johann tried several outdoor climbs during his reckless youth but finally wised up. Now he works out in the gym so he can continue to move his body during the virtual climbs. But he knows that he won't have the strength and agility to keep at it for many more years, and he expects to move into the sedentary "mind-climbing" phase of the sport soon. Johann hasn't climbed with another real human for several years. In fact, Johann hasn't had much contact with other humans since a year or so after he became involved in his pvw. He and his wife, Marta, and their grown children virtually meet every Christmas, but lately they have been at odds over which communal scenario should be the setting for the event. Marta would like a traditional old German winter, while Johann likes the Alpine ski chalet simulation. The children? Their simulations are so strange that Johann can't understand them. They don't seem to relate at all to the natural world Johann knows.

As for Johann's understanding of the real El Capitan, he began to customize his simulation several years ago. At that time he wanted to vary the event structure so he could practice a greater variety of climbing techniques on each ascent. Can Johann see the natural world anymore? If I have trouble seeing the natural world merely after going to the mall, I think it is reasonable to conclude that Johann's understanding of nature has been radically altered. After all, for me nature is more mediated by virtuality than it was for my pre-television forebears; for subsequent, fully virtualized generations, nature will be far more artificially mediated than I can imagine.

And we are already getting gee-whiz glimpses of this future. A *USA Today* feature breathlessly tells us about the CAVE, the Cave Automatic Virtual Environment, under development at the Argonne National Laboratory: "It is astonishing. The system can make computer-animated fish swim around your legs, then rush to nibble cartoon food dropped from a hand-held wand. It can let people thousands of miles apart design a full-scale car together, walk around it, look under the hood. Someday, systems like this may let a grandfather play with his grandchild who lives in another city. Both would interact with full-size, three-dimensional images of the other—like a whole-person video phone call." Wearing only "what look like oversize Ray Ban sunglasses," one can walk through a room completely engaged in three-dimensional VR. As remotely located CAVE sites become linked through high-speed networks, interactions as complex as those of Johann and Austin can begin to be developed. To a population just starting to get used to the Internet, immersive VR will require a further shift in perception. "Says Argonne Labs' Ian Foster, 'We need to get people thinking bigger than they ever have before.' "[8] Start thinking big; the transformation has already begun.

Until recently my life has been virtualized through powerful but blunt mass-communication technologies: radio, television, magazines, newspapers. But now I face a far more powerful engine of individuation and virtualization: networked virtual realities. Moving far beyond the interactivity of telephony and the immersive capabilities of books, radio, television, and movies, NVRs can seduce us completely.

The two key terms here are interactivity and immersion. Johann and Austin are immersed in simulated environments, and they can interact with anyone connected to the network. But Johann and Austin are virtual characters. Pruitt and Barrett have engaged, among other things, their readers' knowledge of virtuality—regardless of whether their readers ever used that word to describe the willing (or unwitting) suspension of disbelief—to conjure up visions of these characters and their cvws.

Technologically there is a chasm between, say, television, on one hand, and fully immersive, broadband networks, on the other. Despite the hype and the energetic visions of the future, CAVEs and CVWs and the like may not be fully realized in our lifetimes. But it doesn't really matter whether scenarios like Pruitt and Barrett's become reality. We are already capable of powerfully immersive, interactive technologies. They may not be as totalizing as CVWs, but they are seductive nonetheless.

Let us examine the primitive (by comparison) combination of the Internet and network television that is operating around the clock today. Already the two technologies are combined in a kind of people's broadband, interactive virtual community development project, as James Barron describes:

> Christopher Fusco watches his favorite television show from a chrome-legged swivel chair he bought at a yard sale. The moment a commercial comes on, he whirls around to his I.B.M.-compatible personal computer, dials up an on-line service and types out messages to other fans who are also tuned in to "The X-Files," the Fox network's New Age answer to "The Twilight Zone" or "The Outer Limits."
>
> . . . Mr. Fusco is the newest kind of couch potato. When not staring at one video screen (his television), he is staring at another (his computer monitor). When the closing credits roll on one, the show is just beginning on the other.[9]

Thousands (soon to be millions) of Fuscoesque Internauts, unable to interact within the virtual reality of the television shows, can at least interact with each other, one step removed from the shows themselves. And the hypemongers are calling discussion groups like this the new virtual communities (for example, the *X-Files* Community, the *Star Trek*

Community, the Wolf Blitzer Community). But online discussion groups are not the only type of seductive NVR. Imagine an *X-Files* MUD; that is, imagine a multi-user dimension—a network environment through which participants all over the world can interact simultaneously—where all those connected are role-playing within the thematic confines of the *X-Files,* or *Star Trek,* or the CNN Newsroom in a post-apocalyptic war zone, or whatever scenarios, rules, and aesthetics you can imagine. Even if it is all merely text-based, even if the only thing you can do is type into your computer and watch your words and the words of others scroll up the screen, you can quickly become entwined in a complex and compelling virtual world.

The most famous such world is LambdaMOO, a virtual house to which participants can connect (or "telnet") via their computers, "where" they can assume online identities, where they meet, communicate, get to know one another, develop social groups, social strata, social structures and policies—in short, where participants create virtual community. And all this is done via relatively primitive computers using only text. Most people seem quite excited by the wondrous possibilities of these networked virtual realities—and in later chapters I discuss these enterprises in more detail—but my point is that networked virtual realities individuate us. They encourage us to ignore, forget, or become blind to our sense of geographic place and community, and they direct our focus toward the self in relation to the mythologies and promises of virtual communities.

The force behind NVRs and behind projections like Pruitt and Barrett's is what Kroker and Weinstein call the "will to virtuality." This evolution into the virtual is promulgated by a "virtual class" of people who benefit from the promotion and development of techno-utopianism, or "technotopia": the complete belief in and acceptance of the "medianet," the digital information highway. The purpose of the medianet is not to enhance communication among individuals but to propel individuals toward virtuality: "The digital superhighway always means its opposite: not an open telematic autoroute for fast circulation across the elec-

tronic galaxy, but an immensely seductive harvesting machine for delivering bodies, culture, and labor to virtualization." This harvest underlies
Kroker and Weinstein's theory of the virtual class: "Cultural accommoda-
tion to technotopia is its goal, political consolidation (around the aims of
the virtual class) its methods, multimedia nervous systems its relay, and
(our) disappearance into pure virtualities its ecstatic destiny."[10]

Given this evolution to virtuality, we must consider what is lost in
our immersion. Let's go back to that moment when I stood beneath the
maple tree on my road. If you allow me, for the sake of argument, the
technological power that Pruitt and Barrett envision, the question becomes, What could be incorporated into a virtual reality simulation of
that experience? I assume that the following elements could be simulated:
the light from the sun and the sky, the tree in all its golden brilliance, the
feel of the cool October breezes, the bird droppings falling onto the road,
the ramshackle house beneath the tree, the sensation of walking toward
the tree and the increasing noise of the birds, the leaves flying in the
air, the road, the view of cars and pickup trucks occasionally passing.

I'll even accept that an amazing variety of contingencies can be virtualized, such as the possibility of birds flying to and from the tree, or the
possibility of getting hit by bird droppings as I stand there in awe. Add to
these all the contingencies that come with a networked system in which
remotely located others are entering the simulation. But virtuality becomes less plausible as local contingencies become more complex, for
example, the possibility of getting killed when one of the pickups driven
by a local teenage drunk runs me over; the possibility that my neighbor
will drive past and wave, stop her car, back up, and tell me that her husband discovered two pieces of my mail left mistakenly in their mailbox;
the possibility that the elderly woman who lives in the house beneath the
tree is heading toward her mailbox to send off the town taxes that she and
I and our neighbors pay to provide crews to plow the road all winter; the
web of relations in town that connect her to me; the difficulties that all
of us encounter in keeping an eye on vacationers' places, in baby-sitting,

in dog-sitting, in picking up garbage dumped on the side of the road by local jerks, in taking care of the kittens dumped on this country road by other jerks; and on and on. In short, try as we might, via the medianet we cannot live a life. The difference between reality and the infinitely realistic irreality of VR, notes Michael Heim, are the biological imperatives of bodies[11] — imperatives, I would add, that include all the biologies of the ecosystem and the vast web of social relations fused to those imperatives. Regardless, we are increasingly attracted by the possibility of transcending our messy, indeterminate, complex physical existence through the technotopia of the net.

In physical communities we are forced to live with people who may differ from us in many ways. But virtual communities offer us the opportunity to construct utopian collectivities — communities of interest, education, tastes, beliefs, and skills. In cyberspace we can remake the world out of an unsettled landscape.[12] The natural frontier has been long since tamed. Leave it behind, say the cybernetic hypemeisters, and settle the wild electronic frontier.

This metaphor conjures up traditional American images of the individual lighting out for the territories, independent and hopeful, to make a life. It fools us into thinking that as natural frontiers become ever more remote from our lives, there is another kind of nature, another kind of wild place, where we can develop and express our human potential. For example, after having considered the Adirondack hiker with the cellular phone, Markoff speculates about future frontiers:

> If wilderness plays an important part in kindling the human spirit, perhaps as it vanishes people are reinventing it in different ways.
>
> In the future with the earth encircled by satellites and everyone wired together by digital links, the new back country may become the world of artificial computer networks known as cyberspace. One can already become lost for hours in the neck of the Internet called the World Wide Web, pointing and clicking a trail through a maze of hypertext documents and digital pictures.[13]

Markoff goes on to quote John Perry Barlow, one of the founders of the Electronic Frontier Foundation, an organization that promotes citizens' rights in cyberspace. "That's the thing about cyberspace," notes Barlow. "It's the last frontier and it will be a permanent frontier. It's infinite and it's continuously changing." Unfortunately, what is silent in our emigration into this so-called frontier is our utter dependence on technology created, provided, and sustained by others. This is a sign not of frontier but of containment, not of our independence but of our domestication.

Humans are the most domesticated of all animals, argues Canadian naturalist John Livingston, because we have become completely dependent on ideas—nurture, technology—over nature.

> There are many visible earmarks of domestication. One, however, must be stressed above all others, and that is the matter of dependence. All domesticated animals depend for their day-to-day survival upon their owners. . . . The human domesticate has become equally dependent, not upon a proprietor, but upon storable, retrievable, transmissible technique. Technology provides us with everything we require. Knowledge of how-to-do-it sustains us utterly. And since none of us knows how to do everything, we are further dependent upon the expertise of countless others to provide even the most basic of daily necessities. . . . Without knowledge of how-to-do-it, or access to someone else who does know how, we are irretrievably helpless.

Networked virtual realities are shining examples of "storable, retrievable, transmissible technique." They are quintessential domesticating engines hiding beneath frontier-like facades. The domesticates of the net are like domesticates in Livingston's view of the natural world: placeless. "Nowhere may the human presence be seen as fully integrated and 'natural' because wherever we may be, or however long we may have been there, we are still domesticates. Domesticates have no ecologic place, and they show it consistently and universally." And in achieving our domestication, we strike a bargain. We get protection from natural enemies while giving up our place in nature. "Domestication confers special gifts, the most important of which is relative freedom from the pressure of natural

selection, meaning at least temporary immunity to many normal eco-
logical constraints," says Livingston. "In return for these gifts, we have
handed back, as it were, the quality of natural, integrated belonging-
ness."[14] The further we surrender ourselves to this ersatz frontier, the
greater our placelessness.

# 2 Immersive Virtualists and Wired Communitarians

Many of the loudest voices of the net tell us that a revolution is going on, and that revolutions mean change, and that this change can be painful for some, but that overall the revolution is good. Notice the emphases, for example, in the following excerpt from National Public Radio's *All Things Considered*, a segment from February 3, 1994, entitled "Pile-Ups Could Be Problem for Information Highway." The real problem, this interview tells us, is the difficulty in getting everyone online. But once online, let the good times roll:

NOAH ADAMS, host: The traffic report for the information network, the Internet highway, is smooth going, but there's a pileup at the America Online ramp causing congestion. The computer on-line services offer bulletin boards and magazine articles, research, and shopping along with access to Internet, the vast grid of computer systems. The on-line services have become very popular, and America Online is the fastest growing. The membership more than doubled last year. But lately the service has often been slow, and sometimes subscribers simply aren't able to log on. Computer analyst Fred Davis joins us from KQED.

After some discussion:

ADAMS: Is it possible that this [electronic communication via Internet] is just a fad, just a fashionable thing to do, you're sort of browsing around and next year we'll be doing something else?

DAVIS: No, this is a major change in the way that humans communicate. And it's bringing people closer together all over the world. People are having international romances over the Information Superhighway, Internet, fax machines, and so forth. It just brings us all closer together, and it's addictive. Once you get the feel of creating a community based on your interests rather than arbitrary geography, it's a really exciting and compelling thing for humans to do.

ADAMS: So we better figure out how to do it.

The net pulls you in—but (apparently) in a good way. It brings people closer together, enabling us to create vibrant communities while overcoming the caprice of geography. It entices us with the promise of romance, love, and (safe) sex. Everybody will get online and communicate with everyone else.

According to this view the old regime is crumbling. The mass media as we know it is dead. Newspapers don't give us news, because in the time frame of the net what newspapers print is old news before ink meets paper. Book publishing is dead when anyone's words can reach millions of potential readers moments after the final draft is finished. The music industry withers away when any musician can make high-quality digital recordings at home and distribute them to the world via the net. Television networks are inconsequential when everyone has the power to produce programming for everyone else. Governments as we know them are increasingly powerless because nations become irrelevant when communication technologies make borders as porous as air. Power goes to those who can control the flow of information. But when all information, all music, all art, all words, all images, all ideas are digitized, then everyone can access, alter, create, and transmit anything, anywhere, anytime. Welcome to the spectacle. Enter the teeming, buzzing cacophony of cyberspace. The revolution is here. The kings are dead. Long live us all— all virtual citizens in the egalitarian, electronic democracy that is the net.

And we participate in this electronic parlor as children of the *convergence:* that moment when the bandwidth widens, when all communication technologies blend seamlessly and transparently into a single, inevitable *pan*media—the melding of telecommunications, computers and computer networks, cellular, cable, and satellite data transmission, television, radio, and print media, and while we're at it, virtual reality technology, biotechnology, artificial intelligence, and nanotechnology, turning every one of us into the liberated, mind-expanded, and globally connected cyborgian citizens of the global community. Says William Mitchell in *City of Bits:* "By this point in the evolution of miniature electronic products, you will have acquired a collection of interchangeable, snap-in organs connected by exonerves. Where these electronic organs interface to your sensory receptors and your muscles, there will be continuous bit-spits across the carbon/silicon gap. And where they bridge to the external digital world, your nervous system will plug into the worldwide digital net. You will have become a modular, reconfigurable, infinitely extensible cyborg."[1]

This vision represents the manifestation of our will to virtuality. If we are to achieve these dreams of the simulacrum, we must know that the virtual can sustain us—that the virtual is life and community. That is, as the virtualization of human relations continues, we come to see its product as life-giving and life-affirming. We recognize that we can live not only a life on the net but a rich and nurturing life on the net (this vision also incorporates the counter-position: we cannot live much of a life off the net). If we accept consciously or unconsciously that the net is life, we accept that it is natural and organic. It is the fertile ground on which we can grow a life with others in our virtual communities.

What is living and what is not is up for grabs in this electronic circus. First of all, you don't have to look hard to find tiny manifestations of the science fiction dreams of the human-like machine. If you start examining any number of magazines about science and technology, you can't swing a dead (computer) mouse without hitting an article about something new in biotechnology and artificial intelligence. Scientists at DuPont, for example, are creating silicon devices that simulate brain functions in increasingly complex ways. As these silicon-based machines evolve, they will work more like organic brains and less like standard digital computers. They will have different capabilities, just as digital machines have capabilities that organic brains don't have. "They will be less like the idiots that digital boxes are now," says Michael Gruber, "utterly dependent on flawless programming, and more like dogs: trainable, but with an inherent set of instincts and abilities, herding our processes and reactions and systems like a border collie runs a flock of sheep." At the Mobile Robot Lab at the Massachusetts Institute of Technology, Rodney Brooks and his team have developed a variety of robots that operate not by a preprogrammed plan but by adapting their capabilities to their surroundings. They read the environment and react accordingly. In short, they learn.[2]

But these machines with seemingly lifelike intelligences pale in comparison to the sophisticated cyborgian organism, the complex human-machine hybrid, already operating within our midst. It has been given different names—I refer to it as the net, while others call it the medianet (Kroker and Weinstein), information (Barlow), the datasphere (Douglas Rushkoff)—but these names refer to the same vast, amorphous, ever changing, immortal being with an unambiguous social agenda: free the individual. Whereas the robotic devices I described may be able to perform tasks that empower humans in a variety of ways, the net—according

datasphere

to many technotopists—is an organic, political entity. It votes libertarian every time.

This libertarian life force can best be illustrated by first examining the Internet. In its current form the Internet is designed to operate as a decentralized network, with multiple pathways along which information can flow. The genesis of this system is described in Howard Rheingold's brief history of the original core of the Internet, the U.S. Defense Department's ARPANET, which was designed so that if any part of the network was disabled (in a nuclear attack), the rest of the network could function and all network traffic could simply be rerouted. There would be no central command. This decentralized web appears organic to observers like Rheingold: "Information can take so many alternative routes when one of the nodes of the network is removed that the Net is almost immortally flexible. It is this flexibility that CMC telecom pioneer John Gilmore referred to when he said, 'The Net interprets censorship as damage and routes around it.'"[3] The Internet thus appears immortal. If you choke off one tendril, information can react and reach its destination via another route.

The question here is one of agency: Who or what can initiate and carry out actions? I considered this question after reading an article by John Perry Barlow in *Wired* magazine—a lengthy piece that attempted to redefine intellectual property in the digital age. I liked Barlow's argument because its premise was that a unit of information is not an objective entity with a fixed meaning that can be transferred from one decoder to another; rather, meaning-making is a collaborative process of negotiation in which participants interpret and construct the meanings of the unit of information in a myriad of ways. Information is a verb, not a noun.[4] What I did not like was that Barlow, while arguing that information is an action and not an entity, still spoke of information as an objective thing throughout the article.

I wrote him a note via e-mail (which he later forwarded to the magazine and which was printed in the June 1994 issue). I told him that I agreed with his overall position but thought that his piece suffered from a "debilitating metaphor":

> Throughout your argument you objectify information while you simulta-neously try to undermine that objectification. That is, you argue effectively that information is not a thing but a process/relationship/verb; however, you cast information as a thing throughout the piece.
>
> I think the danger in how you characterize information is this: Many people are utterly convinced that information is a thing—a commodity that can be transferred—and they base their actions on this flimsy foundation. . . . While you argue against the concept of information transfer, you actually argue against yourself by continually casting information as an agent, not an action. Information doesn't want to be free, informers do. The meanings are rhetorical in nature: They are negotiated, they are constantly constructed and reconstructed during their interactions among participants in the com-municative acts. By saying things like "information wants to change" etc., you give agency to the code—even though your purpose is just the opposite.[5]

My point is simply that the only agents in communication are humans. Things don't communicate; individuals interpret things and give them meaning. The hydraulic tube on my storm door that I discussed at the outset of Chapter 1 doesn't tell me that it is cold outside; I have asso-ciated the temperature and the performance of the hydraulic tube and have given certain performances meaning.

Barlow's response reveals a very different position. He wrote:

> It's a problem of my personal semantics. I call a lot of nonthingish things things, like, to use a big one, this Thing Called Love. Language, or at least English, is limited in its ability to describe the nonspecific action or state of being.
>
> Actually, since I believe information is a life form, there are many cases where information may be seen to act upon . . . uh . . . things.[6]

Ah, so information does indeed want to be free. In Barlow's view, the "thing" has agency, and the net is an organic entity pulsating with infor-mation that seeks its own expression and cannot be censored.

*the
vision* /

But that's not all. According to Douglas Rushkoff, the net is not only alive but has a variety of agendas. Rushkoff describes the datasphere, an entity like Kroker and Weinstein's medianet, which encompasses all electronic communication networks from the vestiges of the traditional mass media to the distributed connectivity of the Internet and of facsimile and telephony technologies. You can recognize the datasphere as the life force it is, according to Rushkoff, if you've grown up after its ascendancy in the past two or three decades. Those who have "don't just receive and digest media. They manipulate it. They play with it. The media is not a mirror—it is an 'other.' They are in a living relationship with it."[7] And this relationship makes possible a symbiotic effort to liberate people and ideas in the face of the monolithic controls over public expression throughout the world.

While most people (those older than about age thirty) who are critical of mass-communication technologies argue that these technologies are used by centers of power to further the status quo, younger people who share the life force with the net, states Rushkoff, have appropriated its reach to unleash "media viruses" to fight the power: "Those who grew up after the development of the datasphere see the media very differently. More than a set of tools, the media is an entity unto itself that must be reckoned with on its own terms. The initiators of media viruses depend on a very optimistic vision of how the web of media nodes can serve to foster new cultural growth. Rather than stunting our natural development by amputating our limbs and numbing our senses, the media can accelerate evolution. The activists . . . believe the media can extend the human, or even the planetary spirit."[8]

Citing the work of Noam Chomsky and others, Rushkoff argues that the datasphere is breaking the hold of the public relations era, the era of one-way mass communication. In the PR era, a few messages came from a few sources, and those messages were geared to maintain the current power structure. Now the datasphere enables interactivity; it fragments information providers. Viruses within the datasphere challenge authority

and provide voice to individuals. And those in opposition, those like myself, do not understand the symbiosis of humans and the datasphere. As
long as we don't look at the new communication technologies as part of
our nature, we can never see them as anything but the enemy of the natural. In Rushkoff's view our relationship with the net is part of what makes
us human; it is our natural world. Our place is in any number of virtual
communities of interest. The new geography—unbound by the nasty,
dull, exclusionary necessities of physical space—is an unending and liberating virtual landscape where we can take social relations and collectives
to a new level.[9]

Some technotopists who believe this argument will admit that virtuality alone cannot sustain a culture, society, or community. But what they
do seem to argue is that the foundation for culture, society, or community is our symbiosis with virtualizing technologies. In *The Virtual Community*, for example, Howard Rheingold tells stories about people who
develop emotional attachments by communicating electronically with
each other in the parenting forum on the WELL, a computer conferencing
system. Some of the most compelling stories involve people who became
acquainted online and who go to great lengths to support each other
in times of crisis—just as neighbors should in the ideal neighborhood.
Rheingold uses these stories to anticipate his critics:

> Many people are alarmed by the very idea of a virtual community, fearing
> that it is another step in the wrong direction, substituting more technologi-
> cal ersatz for yet another natural resource or human freedom. These critics
> often voice their sadness at what people have been reduced to doing in a
> civilization that worships technology, decrying the circumstances that lead
> some people into such pathetically disconnected lives that they prefer to find
> their companions on the other side of a computer screen. There is a seed
> of truth in this fear, for virtual communities require more than words on a
> screen at some point if they intend to be other than ersatz.[10]

His tales of online relationships blossoming into offline human interaction and support apparently validate the foundation that is the online
community. But if you read Rheingold's stories carefully, you begin to

realize that nearly all the participants share a geographic place—a diverse place, undoubtedly, but a place nonetheless: the Bay Area around San Francisco. It is out of this foundation that so-called virtual relationships can grow into something that may not capture but may at least approach real community. It is out of the face-to-face WELL picnics at a public park that simple text on a screen begins to develop into something more than just the image of community.

Rheingold and Rushkoff see the net as a means for social empowerment. I share their goal but fear that both place their faith in a shadow whose images are so lifelike that they appear real. I fear that the continual virtualization of community reveals that geophysical community is dying. As we invest ourselves in the simulation, the simulated phenomena disappear.[11]

The technotopists imbue the net, medianet, datasphere—whatever you want to call it—with agency. They believe that interacting with it is sustaining. They are convinced that it is possible to live within it. Humans lose agency, they say, and the medianet gains it.

I agree that the media virus enables a counterculture. The question is, counter to what? The culture is wholly maintained within the datasphere; nothing else exists. The counterculture that the media virus represents is thus not radical. To place your faith in the empowerment of the datasphere is a quintessentially conservative act: don't worry, be happy; surrender to the power of the datasphere. The most radical and most difficult act would be to resist that power.

But resistance is difficult, because social forces push us to lead virtualized lives. Many of the proponents of virtualization seem to fall into one of two camps. In one camp are those futurists and technotopian visionaries who argue that our destiny is to move from the material world to virtuality, that to examine our evolution is to see the movement away from the body and toward the intellect. To these people, being online is an end in itself; virtuality ultimately must become immersive, thereby making virtual community our goal.

Far less grandiose in vision are those of the other camp, who see the continuing virtualization of everyday life as a way to improve our material lives. Virtuality is a tool to help us solve social, psychological, economic, and environmental problems. To these people, our offline lives define our selves, and our goal is to participate in healthy geophysical communities.

> If we speak of a healthy community, we cannot be speaking of a community that is merely human. We are talking about a neighborhood of humans in a place, plus the place itself: its soil, its water, its air, and all the families and tribes of the nonhuman creatures that belong to it. If the place is well preserved, if its entire membership, natural and human, is present in it, and if the human economy is in practical harmony with the nature of the place, then the community is healthy. . . . A healthy community is sustainable; it is, within reasonable limits, self-sufficient and, within reasonable limits, self-determined — that is, free of tyranny.[12]

It is not unusual to read analyses of our post-industrial, information-infused, media- and image-saturated, transitionary times telling us that traditional communities are irrelevant, endangered, or impotent in the face of sweeping economic and social change.[13] One such vision, put forth by Peter Drucker, represents a commonly held view of the present and near future. In a broad analysis of what he describes as the most pervasive cultural change in history — the rise of the "knowledge society" — Drucker makes it clear that the "old community" is dead. Unfortunately, the social needs fulfilled by those old communities remain.[14]

The knowledge society is one in which the driving economic force is the development and application of new knowledge. In such an economy — one fueled by education and technology — the most important resource is not cheap labor, natural resources, or political will; it is the ability to develop and maintain a culture of learning. Unlike industrial economies, a knowledge society will be able to survive global competition only by providing its members with lifelong education so that they may create and use knowledge productively. "The acquisition and distribution of formal knowledge may come to occupy the place in the politics of the knowledge society which the acquisition and distribution of property and income have occupied in our politics over the two or three centuries that we have come to call the Age of Capitalism" (66). There will be less and less need for individuals to obtain prescribed schooling during a set period (for example, from ages seven to twenty-one). Instead, in order to succeed, individuals will become lifelong students.

Such a scenario depicts communication technologies as the response to a need for knowledge. New communication technologies, remember, can deliver anything, anywhere, anytime (or so we are told). Because knowledge will be widely available, competition among knowledge creators, distributors, and users will increase dramatically. "The knowledge society will inevitably become far more competitive than any society we have yet known—for the simple reason that with knowledge being universally accessible, there will be no excuses for nonperformance. There will be no 'poor' countries. There will only be ignorant countries" (68). And the same applies for entities smaller than countries. Industries, corporations, and individuals will all succeed or fail based on their ability to manipulate knowledge.

At the same time, one of the prerequisites for success in this hyper-competitive environment is constant change, the condition for continual social transience:

> People no longer stay where they were born, either in terms of geography or in terms of social position and status. By definition, a knowledge society is a society of mobility. And all of the social functions of the old communities, whether performed well or poorly (and most were performed very poorly indeed), presupposed that the individual and the family would stay put. But the essence of a knowledge society is mobility in terms of where one lives, mobility in terms of what one does, mobility in terms of one's affiliations. People no longer have roots. People no longer have a neighborhood that controls what their home is like, what they do, and, indeed, what their problems are allowed to be. (74)

The old community may be gone, but our need for the kinds of protection and healing available in the old community has not disappeared. The problems that beset individuals and families—crime, domestic violence, substance abuse, divorce, and so on—will persist and, in an increasingly competitive society like the knowledge meritocracy Drucker describes, will probably intensify. The public sector will not be able to deal with those problems; witness the failures of the welfare state. Nor is the private sector appropriately equipped to handle the "social tasks" of the knowledge society: "In fact, practically all these tasks—whether education or health care; the anomies and diseases of a developed and, especially, a rich society, such as alcohol and drug abuse; or the problems of incompetence and irresponsibility such as those of the underclass in the American city—lie outside the employing institution" (72).

The answer, according to Drucker, lies in the middle ground between the public and private sector. The services and opportunities of the "social sector" must fill the void. "The old communities—family, village, parish, and so on—have all but disappeared in the knowledge society," says Drucker. "Their place has largely been taken by the new unit of social integration, the organization. Where community was fate, organization is voluntary membership. Where community claimed the entire person, organization is a means to a person's end, a tool" (76). The social sector consists primarily of volunteer-based, nonprofit enterprises, from churches to charitable organizations. Through such organizations, individuals can both help and be helped, in a reciprocal spiral that, in Drucker's view, can "create citizenship" and re-create *a sense* of community. That is, only through the social sector—a salve to the socially dysfunctional—can individuals can take part in the process of maintaining a community. "Modern society and modern polity have become so big and complex that citizenship—that is, responsible participation—is no longer possible. All we can do as citizens is to vote once every few years and to pay taxes all the time" (76). Unless, of course, we become engaged in the work of the social sector.

I urge you to feel as sad about this vision as I do. Have we devolved so far that the only way to participate in healing social wounds, creating social connections, and maintaining social bonds is by joining bureaucracies and institutions? No matter how benevolent or socially and spiritually conscious, they are still bureaucracies and institutions. Where is the future of interpersonal bonding unmediated by systems? Where is the future of random, unexpected, unintended, but inevitable community-building of individuals of unlike mind and appearances?

I'm afraid Drucker's social sector—even in its most efficient and grand state—is a paltry substitute for what is lost when geophysical community disappears. The organizations in a social sector unbound by community may serve as important nodes in a social service network, but they are like nodes without the connecting strands. However closely linked one organization is to another, they still float in the ether, islands of service and good intentions. Factor into this mix the pull of the net away from geographic ties and toward the virtual, and you have at best a surrogate community atop the spindly legs of procedural bureaucracies.

So while Drucker correctly recognizes the inadequacy of a society built on the twin towers of the public and private sectors, I disagree that a third support can be constructed to fill the void. Instead, the private and public sectors must be encompassed by the normal functionings of placed communities.

This is precisely the argument that Daniel Kemmis makes in his analysis of why our political system limps along, sometimes failing, sometimes barely fulfilling its charge to develop and maintain a good society. "If we cannot turn the clock back to some imagined golden age," notes Kemmis, "if we choose not to hurtle into a future which destroys most of what makes life worth living in the region, and if our current politics of polarization is steadily undermining both our public life and our capacity to shape a viable economy, then we need to try to envision a fourth way . . . an alternative which I have spoken of as the 'politics of inhabitation.' "[15]

Kemmis argues that any consideration of reinvigorating political structures and processes that is not tied to place weakens the effort. What has been lost is the vibrancy of republicanism with a small *r*—the way of life

GAP:
w/r
where
is
the
dialectic
tension?

in which individuals are bound to others through desire and necessity to work together, despite their differences, because they all have a stake in the betterment of their common place. "Republicanism (in the early, formative stages of the nation) was an intensive brand of politics; it was, heart and soul, a politics of engagement. It depended first upon people being deeply engaged with one another . . . and second upon citizens being directly and profoundly engaged with working out the solutions to public problems, by formulating and enacting the 'common good' " (12). But, Kemmis shows, this ideal form of collective action had quite a limited tenure in the United States. As politics and government evolved, domestic tranquillity came to be ensured not through the public deliberations of the entire community but through a legal structure, a complex procedural system of checks and balances. This structure gained precedence over republicanism because of the fear of what James Madison called the tyranny of the majority. Therefore, the political structure had to evolve to ensure private rights and keep competing individuals separated in public life: the reigning mentality became "I don't care what they do in their private lives just as long as they don't impinge on my rights." As long as there was more territory to expand into and as long as there was equal protection under the law, the citizens did not need to confront their differences in order to solve their problems.

> Republicans believed that public life was essentially a matter of the common choosing and will of a common world—the 'common unity' (or community). . . . The federalists argued that it was possible—in fact it was preferable—to carry on the most important public tasks without any such common willing of a common world. Individuals would pursue their private ends, and the structure of government would balance those pursuits so cleverly that the highest good would emerge without anyone having bothered to will its existence. It was no accident that this approach to public life was put forward by people who were centrally interested in creating optimal conditions for an expanding commercial and industrial economy. (15)

What we are left with is a politics of opposition between, on one hand, proponents of individual freedom and, on the other, bureaucracies that regulate individual actions. The missing middle ground cannot arise without a collectivity of individuals committed to inhabiting a place and to enduring and working with each other to improve the condition of everyone in that place. "We have largely lost the sense that our capacity to live well in a place might depend upon our ability to relate to neighbors (especially neighbors with a different life-style) on the basis of shared habits of behavior" (79). We have come to accept that committing to a place is an expression of personal choice and personal taste. Or, quite often, it is forced on us either by our need to find work or by our need to live with or near another person.

I have a couple of friends who want to move away from northern New York. Our north country is a vast stretch of farms, rolling hills, and rivers dropping out of the Adirondacks to the south and flowing into the great St. Lawrence River to the north. It is poor, cold, and, to many observers, desolate. My friends tell me that it can no longer sustain them. One is unhappy with her job and says, like many who talk about leaving, that she wants to live somewhere warmer. My other friend, however, loves the cold and the snow but wants to move east of the big lake, Champlain, and settle in Vermont. He believes that his business—often on the brink of bankruptcy here—has a better chance over there. But that is only part of the issue. Everything, he tells me, is better over there: the economy, the social life, the environment. People take better care of their homes and their businesses. They care more about their culture; they are less provincial; they aren't always suspicious of change and have achieved a balance between accepting the progressive and holding on to traditions. Vermont, he says, is more attuned to individual initiative. Simply put, there's just a different ethic over there.

I listen to these complaints and say, "Well, yeah, you're right. Life in the north country can be difficult . . ." At this point I can't really think of much to say besides "but I want to stay." I'll admit I came here to take a job—but I also came here because I thought it might be a place to commit to. I'm willing to inhabit a place with faults, a place that is difficult to live in, a place where I cannot satisfy all my wants and needs. I am willing to do this for the sake of becoming part of a community.

My friend who wants to move to Vermont wants to commit to that place. That is good. I have been in his position; I've given up on a place and moved away from people I was close to. Nonetheless, I am disheartened when I hear him talk about moving. It is such a long process to develop a friendship; I invest so much time and effort, and the older I get, the harder it is to stay in touch with remote friends and family members. No amount of telephone conversations and e-mail can re-create the experience of sharing daily life. So when I hear people close to me talk of leaving, part of me wants to dispense with them immediately and go nurture other local relationships. Of course, I cannot do this; one doesn't eliminate emotional ties so coldly. No, I act as if I believe that after my friends or family members move, we can maintain our relationships by keeping in touch via technology. But we cannot really maintain those relationships, because over time the foundation for our relationship, the social and geographic ties of a common place, fades.

Sometimes I find myself assessing others on the basis of their interests in staying here. This is probably a foolish criterion. "The longing to become an inhabitant rather than a drifter," says Scott Russell Sanders, "sets me against the current of my culture, which nudges everyone into motion. Newton taught us that a body at rest tends to stay at rest, unless acted on by an outside force. We are acted on ceaselessly by outside forces—advertising, movies, magazines, speeches—and also by the inner force of biology." [16]

In Chapter 1 I illustrated how I found moments of placed community via the (movie) screen and how my ability to see the natural world is increasingly mediated by the (television or advertising) screen. Baudrillard describes the second condition as a prison, likening it to a science fiction story in which a group of people are suddenly glassed off from the rest of the planet.[17] Bill McKibben, in his withering analysis of television and its culture, describes the effects of screens in a similar way. But McKibben offers us hope. He seems to believe that we can turn off the screen and, after a while, regain a closer view of, and a stronger connection to, our local life—a connection that will help us value, preserve, and possibly restore the integrity to the places we find ourselves.[18]

I am persuaded by Baudrillard but want to act with McKibben; that is, I find myself in the maddening position of recognizing the ever-increasing artificial mediation of the natural world while remaining committed to breaking down that mediation to regain a sense of the here and now. I'm seeking what Albert Borgmann describes in *Crossing the Postmodern Divide* as postmodern realism, the necessary response to our hypermediated condition: "Having left modernism behind us, we now have to decide whether to proceed on the endless and joyless plain of hypermodernism or to cross over to another and more real world. For this country in particular, the latter task comes to settling down in the land that has come to be ours, to give up the restless search for a hyperreal elsewhere, and to come to terms with nature and tradition in a patient and vigorous way." As we are on the brink of surrendering to the most powerful mediation engines of all, we must resist the lure of the hyperreal. And if we must become wired, we must turn that telepresence toward, as Borgmann describes it, the "focal realities" of the local places we inhabit.[19]

One day I received an e-mail message directed to all the members of an electronic forum—a "list"—devoted to discussing community computer networks. The writer described a murder that had occurred in her community: a child had been shot and killed by another child. She wondered what those of us interested in community networks had to say about that. Two responses follow.

From: MILTON LOPES <MLOPES@UGA.CC.UGA.EDU>
Subject: Re: A shooting in the community . . . or how do we stop the hemorrhage

X-To: "Communet: Community and Civic Network Discussion List" communet@elk.uvm.edu

Teenage violence is a subject that needs to be discussed on a national forum. It involves no less than our future as a nation. If this is not the place to discuss it, where is the place? I recently facilitated a community meeting following the senseless shooting of two students by a fellow student who they had earlier been harassing. The community was in an uproar. Attending were school officials, police officers, community leaders, parents, and youth. The youth made more sense than any of the other groups. They simply called for teachers who cared, parents who spent some time with them, complete families, the teaching of and practice of morals, in the marketplace, at home, in school, in government. They looked at us adults, and found us long on verbiage, but short on example. I have facilitated other meetings in which adult posturing was simply no answer to the pleas of the children. We have abandoned the basic principles which for most of us were taught by our elders. We are too busy consuming, and selling, and living for the moment. Listen to the children. I hope your grief finds resonance with this entire list.

Milton E. Lopes

From: Tres English <tenglish@WEST.CSCWC.PIMA.EDU>
Subject: Making more time (Was: A shooting in the community . . . )
[After quoting from the preceding message, English added the following.]

I have concluded that the real reason for this accelerating breakup of our society are those things which physically break us up on a day-to-day, moment-to-moment basis. That is the road system, TV, financial system, etc. that have the effect of requiring us to spend more and more *time* separated from family and neighbors.

It is not quality time that matters. It is simply time.

***I have a question for participants in this list.***

How can we structure a community computer network so that we are able to recreate more stable geographic communities where people have more time to spend with each other, both as families and as neighborhoods?

I don't think there is a solution to kids killing each other, and all the rest, unless we can recreate the foundations of a stable society—the continuing, unplanned interactions between the same people for a long period of time.

Tres English.

Tenglish@west.cscwc.pima.edu

A community is bound by place, which always includes complex social and environmental necessities. It is not something you can easily join. You can't subscribe to a community as you subscribe to a discussion group on the net. It must be lived. It is entwined, contradictory, and involves all our senses. It involves the "continuing, unplanned interactions between the same people for a long period of time." Unfortunately, communities across the nation are being undermined and destroyed by a variety of forces. Global computer networks like the Internet, for example, represent a step in the continual virtualization of human relations. The hope that the incredible powers of global computer networks can create new virtual communities, more useful and healthier than the old geographic ones, is thus misplaced. The net seduces us and further removes us from our localities—unless we take charge of it with specific, community-based, local agendas. These agendas are currently under development in many communities through the community network movement. If we do not, as communities, as a society, support this movement, we risk the further disappearance of local communities within globalized virtual collectives of alienated and entertained individuals.

# Part Two

## The Globalized

## Individual

Just machines to make big decisions

Programmed by fellows with compassion
and vision

We'll be clean when our work is done

We'll be eternally free, yes, and eternally
young

What a wonderful world this will be

What a glorious time to be free

What a wonderful world this will be

What a glorious time to be free

—*Donald Fagen, "I.G.Y.*

*(What a Beautiful World)"*

## 3 Virtual Vermont: The Rise of the Global and the Decline of the Local

It's fall. You are driving along the back roads of Vermont, winding around mountains and through colonial villages amid gusts of copper leaves. As you descend into an open river valley, you see the patchwork of small farms and silos and the occasional church steeple, just like the Vermont pictured in the brochures sent by the state tourism bureau. Up ahead you spot a farm stand, and you slow down. The stand is an old wooden structure, solidly built but listing a bit to the northeast, battered from years of winter storms whipping up this valley from the south. To the right of the structure a few rows of dried corn stalks—some still intact, some broken and toppled—are brittle evidence of the past summer's crop. To the left sits a white wooden clapboard house, and beyond that, several faded red barns. Just past the house lies a leafy pumpkin patch. Already a farmhand has been harvesting orange pumpkins and laying them in rows across the front lawn of the house. Another couple of rows line the front of the stand. You pull off the road and get out of your car.

Three thousand miles away, a woman in Oregon sits in front of her computer screen. She is exploring the World Wide Web through her Mosaic-style interface, which allows her to follow a path of never-ending hypertext links. But in this case she is armed with an address—http://www.cybermalls.com—for an online business service called Cyber-Malls. Once connected, she can point and click her way to a number of businesses organized in groups. After scanning the possibilities, she chooses *CyberMont,* and a page appears with the following heading:

CyberMalls' CyberMont Home Page
Welcome to CyberMont, CyberMalls' showcase of Vermont Products and Services.

She scans the options in CyberMont and chooses *Storefronts and other Products.* When she is presented with a page-long list of hypertext links to Vermont companies and services, she chooses *Cold Hollow Cider Mill,* where she can follow the company's links to see a variety of its products and ordering information. The first link she takes introduces her to the company:

Welcome to the Cold Hollow Cider Mill
Dear Friends,

   Fall in New England is a glorious time of the year. The spectacular color of the leaves, the cool crisp fresh air, and the harvesting of summer crops all combine to create the look and feel of times gone by. Here in Vermont we cherish old-fashioned values and purity of products. The Cold Hollow Cider Mill specializes in the finest Vermont foods, made with the freshest ingredients.

   We invite you to enjoy this on-line catalog and fill your pantry with an assortment of apple products, maple syrup, jams, jellies and much more. We also offer a wide selection of gift packs featuring these same fine products. If you have any questions, or would like some advice on putting together your own combination, please do not hesitate to call us. Our staff is available daily from 8:00 A.M. until 6:00 P.M. EST for your convenience. Please take advantage of our toll-free 1-800-327-7537 number. On-line ordering is available through our CyberShop Menu.

Remember, it's not too early to start thinking about your holiday needs.
Sincerely,
Eric & Francine Chittenden

After exploring all of Cold Hollow's links, she decides to purchase a box of Vermont Macintosh apples grown in the Champlain Valley. Later she checks out a Vermont health food store that sells premium apricots from Turkey. She pays for it all with Ecash, an electronic form of currency that enables her to use credit without revealing her Visa card number.

While those transactions are spanning the continent, you bite into one of the apples you've just purchased at the farm stand. You love the taste, the snap and tang of this local Macintosh apple. But you probably don't realize that the apple was grown thousands of miles from Vermont. Indeed, if you had stopped here and bought spinach on May 1 or lettuce in early June, cucumbers and corn before July 4, cantaloupes in mid-July, or grapes, oranges, and peaches anytime, you would not have bought produce grown on this farm. In fact, you would have bought produce grown neither in Vermont nor in New England. You might have had some New Jersey corn and some South Carolina peaches, but it is far more likely that everything you would have bought would have come from California or Florida.

As you stand by the side of the road eating your apple, you are consuming not Vermont produce but Vermont. Where the apple comes from doesn't matter. What is grown on this farm, if anything, doesn't matter. All that matters is that you see, feel, hear, and taste something that seems to be Vermont. Likewise, the customer in Oregon who buys Vermont apples or Vermont health food, fudge, smoked meat, and so on, is purchasing the image of Vermont simplicity and rectitude—the clean, rural, solid, and unspoiled feel of Made-in-Vermont. And this thing, this Vermont, the legacy of a way of life long since passed, is constructed in ever more powerful and pervasive ways by electronic communication technologies.

Food distribution in Vermont is an example of globalizing opportunities enabled by advanced electronic communications. The result is the creation of an image of Vermont constructed on a system that supports mass-produced, centralized food production and—notwithstanding Cold Hill Farm and others that sell produce grown in Vermont—inhibits local food production.

There are three kinds of farm stands in Vermont. In the first, the farmers sell only what they grow, either at stands adjacent to their farms or out of the backs of trucks parked along a road. Fewer and fewer farmers do this. Many more buy produce from a Vermont produce distributor so that they can begin selling a variety of attractive fruits and vegetables as soon as the spring weather allows. As their growing seasons progress, these farmers will replace remotely grown produce with the harvests from their own fields, so, for example, by the third week in July they may stop selling New Jersey sweet corn and start selling their own.

But many farmers are discovering that it doesn't pay to grow their own produce. They can sell their own corn for, say, eight to ten dollars a bushel after putting seven dollars' worth of work into the production of that bushel; or they can buy remotely grown corn from the distributor for seven dollars, put no money into production, and still sell each bushel for eight to ten dollars. Further, when locally grown produce is plentiful, the farmers' ability to profit by growing their own produce drops precipitously. In August they can buy remotely grown tomatoes, summer squash, zucchini, and peppers for less than it costs to grow them.

The smart farmers buy all their produce from the distributors, leaving their farms as a front. One farmer, for example, cut back his production and upped his purchases year by year, and now he merely grows a few rows of corn and some pumpkins by the road. The rest of his many acres are mowed. Another farmer asked the distributors not to deliver during the day so that his front would not be compromised. These places are no longer farms; they just simulate farms.

How can it be more profitable to buy wholesale produce grown thousands of miles away than to grow your own and sell it next to the fields in which it grew? Most people would assume that this is the result of developments in farming science and technologies and of changes in government regulations and subsidies. But the key to this system is the revolution in electronic communication technologies. Local food distributors, such as those in Vermont, are linked through communication networks with independent food brokers, the big corporate growers, and the large trucking firms to bring the centralized, mass-produced food to local markets.

The brokers lie at the heart of this network. They buy produce from the big growers to fill the orders that come in from the local distributors. They also select the trucking firms to carry the orders to their destinations. The brokers are constantly in cellular connection to all parties— the growers, the truckers, and, in this case, the Vermont distributors— as they go from the fields to the warehouses to their offices. Further, they move their operations frequently (depending on what produce is being harvested where), so their business is completely enabled by mobile cellular communications.

The trucking companies can track their trucks via satellite to ensure that each load reaches its destination as efficiently as possible. The satellite connection also makes possible two-way communication between truckers and dispatchers to monitor the status of the shipments and handle problems. The Vermont distributor can plan its warehousing and distribution by tracking the continual influx of out-of-state trucks carrying produce that will be sold and shipped to purchasers throughout Vermont.

The key to this dynamic is for all parties to continue to increase the efficiency of the distribution system: from the brokers' cellular connections to the truckers' route and load management to the distributors' delivery schedules, the entire process is a complex communication system in which knowledge about the current and future states of the produce, the market, and the transportation are analyzed and acted upon.

As the system is fine-tuned, the local Vermont farmers have more difficulty competing, and most find it impossible. Some Vermont food distributors have adopted a buy-Vermont-first policy to help keep local farmers in business. But trading with twenty-five Vermont producers or farmers on a given summer's day is less efficient than dealing with the two or three alliances of big growers, brokers, and truckers. So the state depends on the will of the Vermont food distributors to purchase whatever local produce is still available in order to keep Vermont looking and feeling like Vermont. This is a virtualization on a grand scale. Rural Vermont is undermined, but it is the rural image that drives the state economy. If Vermont loses that image, what is Vermont?

This question was raised in June 1993 when the National Trust for Historic Preservation released a report that put Vermont on its list of the eleven most endangered historic places. The Trust attempted to publicize the threat that commercial and technological development posed to the state's image.

This tension between the supposedly real Vermont and its image was treated on an episode of the television show *Good Morning America* on April 26, 1994. The report featured the work of a well-known Vermont artist whose prints often depict Vermont buildings and landscapes. This is art rooted in a sense of place, said the reporter, who was shown driving his car through the countryside and who assured us of his authentic knowledge of Vermont by noting that he went to college at Dartmouth, just over the New Hampshire border. In one segment of his report, the camera would show a picturesque shot of, say, a church—an image designed to make a viewer think "That's a classic New England scene"—and then would dissolve into the artist's conception of that scene. The dissolve showed how seemingly faithful the artist was to the place—how the artwork represented a real place and thus was somehow authentic.

But of course that artwork is not a photograph; the medium, a type of printing technology, enables the representation to take on an idealized form. The lines are crisp, the colors bright, the shadows straight. The simplicity and rectitude of the church is shown clearly in the print. But based on *Good Morning America,* you might say that the "real place," the image you see on the screen, had that same simplicity and rectitude.

The trick is that the video image is also an idealized construction. In neither the artist's view nor the video camera's do we see the Mobil Mini Mart on one corner of the adjacent crossroad. In neither view do we see the condominiums built in a faux New England style on the side of the nearby mountain. But those developments are not hidden. Indeed, they are what fuels concerns like those expressed by the National Trust for Historic Preservation. And people who agree with such concerns usually respond by voting to enact strict environmental and historic preservation standards that regulate development and land use. Vermont has many such measures in place, and yet the quality of its Vermontness is declining. Local life is actually being undermined by means far more insidious than development of strip malls and condos. Indeed, few of us understand how vast technological-economic engines fueled by electronic communication networks replace local life with the image of local life.

More important, we forget about the hooks that pull us in and bind us to the globalized life. As a typical consumer, I have come to expect fresh vegetables to be available year-round. I think I am healthier and I know I'm happier always having a range of vegetables to consume. If my local food stores took away my grapes or red and yellow peppers in January, I would be one of the many complainers lined up at the courtesy desk. And yet I live in a remote region with a short growing season.

Just as local food production has been supplanted by mass food production, local culture and community are being supplanted by mass culture and the image of virtual community. That hook has been embedded since the rise of the early mass-communication technologies and is now nearly impossible to remove. Not only do I need sweet seedless red grapes from Chile all year, I also need a never-ending stream of increasingly compelling entertainment and intellectual stimulation, much of which distracts me further from my local social and physical world.

My distraction has a variety of sources, each of which has been offered as the answer to the question, What is, or will be, the Information Super-highway? Some say that this Infobahn is the Internet, the vast computer network of networks that we gain access to through companies like Net-com Communications, Performance Systems International, and UUnet. These companies sell such services as the ability to connect to remote computer servers around the world (through telnet), to search and re-trieve information stored at remote sites (through file transfer protocol, gopher, or the World Wide Web), to send and receive e-mail, to engage in synchronous chat with others around the world (through MUDs or Inter-net Relay Chat), and to participate in asynchronous discussion forums (such as Usenet or listservs).

Others will say that the information highway is composed of both the Internet and the networks supplied by what I like to call the BANC—the Big American Net Companies, like America Online, Prodigy, Compu-Serve, and the Microsoft Network. Although these companies offer access to the Internet, their primary service has been to create microcosms of the Internet, each with a national reach. That is, a subscriber to one of these networks can communicate with other subscribers around the nation through asynchronous discussion forums and synchronous chat sessions, and can conduct research through information search-and-retrieval ser-vices.

But the most likely manifestation of the information highway will be whatever network is created by the broadband communications giants. These are the evolving mega-corporations that control an ever-shifting combination of media among telephony, cable and broadcast television networks, movie and television studios, cellular and satellite communica-tions, and computer hardware and software. (The communications giants for the most part encompass the Internet access providers and the BANC, but they do not, nor does anyone else, control the Internet.) Add to this mix a myriad of small and large companies engaged in virtual reality development, CD-ROM companies, and the computer game industry, dominated by giants like Sega and Nintendo. Collectively, these entities will offer us a range of products and services from the much-ballyhooed and little-needed movies-on-demand to interactive multimedia networks offering games, home shopping, and information retrieval to immersive networked VR geared primarily toward entertainment and advertising.

The source of my distraction—and the essence of the information infrastructure—is, most simply, all these entities, all at once, all the time. These are the alliances of growers, brokers, truckers, and distributors that produce, sell, and deliver mass culture. Their collective force sustains us.

Within this environment the most powerful distraction engines are networks in any of their forms. And the most compelling dynamic within any network is the individual interacting with others. Networks currently offer primarily text-based interactions, but as they evolve toward multimedia and immersive VR (recall the Personal Virtual Workspaces), their attraction and our distraction grow.

The networks present the attractive possibility of finding new places that transcend the mundane physical spaces restricting our daily lives. We therefore become concerned with developing and maintaining remote, disembodied relationships. Says Stephen M. Case, the founder of America Online, "Our goal is to build a strong sense of community. . . . People are more mobile than they used to be. The vision 50 years ago in which you grew up in a town and spent your life there and knew everybody is now the exception, not the rule. I think people have a thirst for community. Interactive technology is not the be-all and end-all. But in a small way a service like AOL can help bring people together to discuss issues they care about. It's a core part of this medium and something, I think, that the larger companies looking to get into the market don't appreciate. It's not just about video-on-demand and interactive shopping."[1] According to Case, you need not be concerned about reintegrating your life into your surroundings, you need only subscribe to any of a number of national or global communities.

But what kind of community is possible on such a large scale? The entities that Case and others label virtual communities resemble entities we all encounter in our offline lives. They do not constitute communities—online or offline—but they often feel like communities. Robert Bellah and his colleagues have called these collectives "lifestyle enclaves." [2]

In *Habits of the Heart* Bellah and his co-authors note that *community* has many meanings, and the term is often used in connection with a way of life—the skiing community, the show dog community, the criminal justice community, the truck driver community. But a true community is a collective (evolving and dynamic) in which the public and private lives of its members are moving toward interdependency regardless of the significant differences among those members. In contrast, lifestyle enclaves are segmental because they describe only parts of their members' private lives—usually their behaviors of leisure and consumption—and celebrate the "narcissism of similarity" through the common lifestyles of their members. Anyone different is "irrelevant or invisible in terms of one's own lifestyle enclave." [3]

Lifestyle enclaves flourish where individuals need not depend on others for much beyond companionship in their leisure lives. As individuals rely more on national and global ties than on local ties, the need for complex, integrated communities—collectivities of interdependent public and private lives—is replaced by the need for isolated individuals to bond through lifestyle enclaves, which provide only the *sense* of community.

This is my life. Think back for a moment to the ephemeral moment of social bonding that I experienced at the local movie theater. Look at the anecdotal evidence I used to lay claim to community. I said I saw "many familiar faces: college students and faculty, some local joggers—people I knew from Frozen Foote, a series of winter road races in a neighboring town—and a few members of the local bicycle club and the local chapter of the Adirondack Mountain Club." To me that group was the convergence of different enclaves of interest in which I participate. To someone else that group could have been configured very differently. What's worse, many people may have considered that group not as a configuration at all but as just a collection of individuals out to the movies. Why these diverse possible viewpoints? Because despite the way I want to interpret that event, there was no *public* in that theater; there was no gathering of a people with a common interdependence based on the social, psychological, economic, and environmental necessities of a particular place and time. We were merely an aggregation of individual consumers.[4]

Our public lives are shaped by economies that focus our attentions on national and global matters. The regional economy where I live, for example, is dominated by four colleges, several multinational manufacturers, logging, tourism, and farming. Farming—as in Vermont—is in decline, and the remaining industries focus nearly all their energy on competing on a national and global scale. The locally owned retail businesses have gradually disappeared after losing their customers to the national chains in outlying shopping centers. The universities require their employees, from custodians to presidents, to work toward two goals: to develop wide-scale (regional, national, international) renown for the institution, and to provide education for the students, nearly all of whom come here for four years and then return to homes all over the world. (Add to that the continual growth in distance learning that I discuss in Chapter 8—a trend that will globalize not only higher education but also the local grade schools.) The manufacturers and loggers can survive only by exceeding industry standards established in a global market, while the regional tourism industry is manufacturing and broadcasting images designed to persuade consumers worldwide to come and interact with those images for a week's vacation. For most of us in this economy our public lives thus concern issues external to our localities.

Further, one kind of local public space has largely disappeared: the places for public discourse about the issues, problems, and celebrations that affect a locality as a whole. Life in American cities and suburbs is notorious for having eliminated public space for public discourse. When we combine the belief that our economies have relieved us of the need to rely on our neighbors for our economic well-being, the hostile environments that plague our cities and isolate our suburbs, and decades of

urban and suburban planning that have inhibited the creation and maintenance of spaces in which individuals can easily interact, we come to
believe that public life beyond work is either nonexistent or so remote it
is irrelevant to daily life. The planners, as noted by Ray Oldenburg in *The
Great Good Place,* rarely consider such public spaces—"third places," he
calls them—important anymore.[5] In spite of this, however, the romantic
notion persists that small-town life includes public discourse and engenders an interdependency among co-located townspeople. This way of life,
I'm afraid, is lost, too.

To me this loss is best illustrated by the charade that journalists and
politicians play every election cycle when they go in search of "real
people" congregating naturally in public places throughout small-town
America. In cities journalists can stand on streets and shove microphones
into the faces of harried, vaguely fearful, or resentful passersby. In the
suburbs the only thing that passes for public space is the shopping mall—
hardly a site of local public discourse on the local public good. But where
in rural America, in which public space supposedly still exists, can journalists go to find a public? They go to one of the only remaining icons
of public gathering that seems not wholly compromised by ideology or
some special-interest group. They go to the coffee shop.

But, of course, there is no public there. Reporters and politicians may
find a slice of the local populace—retired men, the unemployed, the few
locals left who actually work on Main Street (all of whom typically fall
on the conservative end of the political spectrum)—but that locale is no
more likely to reveal a public than would a Catholic Church service, a
Masonic Temple meeting, a Macintosh user's group, or a meeting of a regional chapter of the People for the Ethical Treatment of Animals. Still,
politicians go to the coffee shop to appear to be in touch with real people;
journalists go there either because they're lazy or because they realize they
can probably obtain quotes from people who speak with regional accents
or who don't use proper grammar—both of which provide authenticity

to the journalists' reports. The entire enterprise is futile, because the heart of a community is no longer located anywhere.

If reporters or politicians came looking for what passes as public space in my town, they would have to interact with locals' personal lives. Their best chance of garnering a cross-section of the town would be to spend a weekend between November and March shivering in one of our three ice arenas to watch hours of youth, high school, and college hockey and figure skating.

Recently my town was the site for what was billed as a community celebration to honor a local woman who had won a gold medal in the Winter Olympics. School children were bussed to one of the large college hockey arenas. Town officials, religious leaders, college administrators, state and federal representatives, and the "general public" filled the arena for what turned out to be a pleasant event of great goodwill. Certainly, the Olympic athlete deserved congratulations for her outstanding performance under the kind of intense pressure that few of us ever have to experience. Those who organized the event were right to do it and did a fine job. Unfortunately, I felt at this event something like the tourist at the Vermont farm stand consuming the image of Vermont.

The Olympian had moved to the area a few years before and as a result of the demands of training and performing had not spent a lot of time actually living here (although she had devoted much of that time to community service in the schools). During the ceremony we were shown excerpts from a CBS television interview with the Olympian, and later we witnessed the unveiling of a mixed-media rendering of the athlete in action, a piece done by an artist from another part of the country who was creating a series of Olympic art and who had shipped the piece north just for this event. Several speakers said that they hoped the Olympian had plans to stay in the area and make it her home. We all engaged in the images of community celebration, but the event felt at best like a genuinely affectionate welcome to and celebration of a relative newcomer from a lot of people who really didn't know each other (many of whom, like myself and other college people, had not lived here all that long themselves), and at worst it felt like a play about community bonding.

With such large-scale events rare, the closest things we have to public discourse are meetings of the school board and the town board, but those are primarily the sites of special-interest debates: somebody wants a tax variance to build a new hydro-dam; an informal group of parents wants to take action regarding a new school bus schedule. Although such deliberation is important, it rarely involves a common public debating the common good.

Here, as in most places, our sense of community arises out of interest-group interactions like these. Local community is the aggregation of enclaves of interest and occurs, most significantly, where those enclaves intersect. And accordingly, much of what we may think of as public discourse occurs in our private lives. Few clear lines between public and private exist in such communities.[6]

Overall, the largest area of intersection among enclaves in my town involves child-rearing: school functions, the Parent-Teacher Association, school sports, school board meetings, community youth associations, and other mostly leisure- and consumer-related activities. Recognizing this, one town in the region has raised a sign near the main road bearing a version of an African cultural proverb: "It takes a whole village to raise a child." This is a powerful idea. But in practice, what it winds up meaning is that it takes an active and well-rounded mix of interacting co-located interest groups to raise a child.

An entity we call a community is sustainable when it comprises a critical mass of interest groups tending to its social, psychological, economic, and environmental needs. Because most of the work of a community does not occur in public space, it must occur through this range of interest enclaves. Communities are in danger when the viability and range of those groups are undermined.

The net can either enhance communities by enabling a new kind of local public space or it can undermine communities by pulling people away from local enclaves and toward global, virtual ones. The second trend is in ascendancy. Much of the net is a Byzantine amalgamation of fragmented, isolating, solipsistic enclaves of interest based on a collectivity of assent (regardless of the minor dissents—"flames"—that occur within them; indeed, the much-discussed phenomenon of flaming is merely a symptom of the lack of real community amid this impoverished thing called virtual community).

Examine again the words of Stephen Case: through the net people can build a "strong sense of community" because community based on place is gone. Come with me, he says, and find online what we've lost everywhere else.

## 4 Seeking Public Space in a Virtual World

In the fall of 1993 I found what I thought was a new kind of public space. It was a compelling online enterprise through which I could interact with others who shared professional interests involving electronic communication networks and cyberspace. But it seemed to be more than that. In this place I could communicate in a relaxed and sometimes playful way about the things we all were studying and the issues we all cared about—and I could do it at any time of the day, any day of the week. It was a virtual parlor that never closed. What's more, once I became an official member, I could create my own private place within this new agora. What I had discovered was MediaMOO, a specific kind of multi-user dimension (MUD) for people interested in the intersection of media and cyberspace. All it took to begin this process was to log on to my Internet account and type "telnet purple-crayon.media.mit.edu 8888," and I was connected, first as a guest and later, after applying for and receiving membership, as "sdf"—the name I gave my "character."

According to its principal creator, Amy Bruckman (along with her colleague Mitchel Resnick of the MIT Media Lab), MediaMOO is a "text-based, networked virtual reality environment" designed to serve as an ongoing professional meeting place for media researchers.[1] That is, MediaMOO is like the lobby, bar, and hallways of a hotel where a convention of aficionados of new media is under way, where people meet, talk, and schmooze.

Such claims about the potential of MediaMOO have been repeated in national publications. Howard Rheingold reported in *Wired:* "In MediaMOO, as in any scientific conference, you can look at other participants' badges and see what they have to say about their special interest. People can find themselves in a hallway or a room with a group of strangers, look at their virtual badges, and strike up conversation." MediaMOO, wrote David Bennahum in *Lingua Franca,* is one of the few "serious MOOs used mainly by researchers to share information, network, and exchange job tips. Job listings cover their virtual bulletin boards, electronic greeting cards are swapped, research ideas shared. Unlike with e-mail, there is an element of serendipity to MOOs—maybe you'll run into someone unexpected, with useful information."[2]

But a far more powerful claim made about MediaMOO is Bruckman and Resnick's comparison of MOOs and MUDs to the third places described by Ray Oldenburg. To those who consider it a third place, MediaMOO is far less formal than a hotel lobby at a convention. "In a very real sense," agrees MIT Media Lab Founding Director Nicholas Negroponte, "MUDs and MOOs are a third place, not home and not work." In this view the MOO is the coffee shop or neighborhood pub where the community—in this case the community of media researchers—comes to hang out. Bruckman and Resnick, referring to the results of a survey of members, note, "A number of [MediaMOO] users commented that their most meaningful interactions on the MUD were with the 'regulars'—the people who use MediaMOO the most and are most likely to be logged on at any given time. Oldenburg emphasizes the importance of 'the regulars' to a third place—they give a place its character."[3] So not only is MediaMOO a "place," but it is one of some depth; it has regulars who bring with them history and experience; it has character, a kind of communally perceptible look and feel.

Before I explore whether MOOs can function as a third place, a glimpse of the complexity of MediaMOO will clarify how participating in it becomes such a compelling act. The description that follows is not the result of systematic research on my part but more a collection of musings and memories I've culled from my experiences "at" MediaMOO. The evidence I offer to support any claims I make about the nature and functions of MOOs should be understood in those terms.

When I first began to log on to MediaMOO, I, like all other newcomers (called "newbies"), would see the following words scroll up my screen as soon as I connected:

The LEGO Closet
It's dark in here, and there are little crunchy plastic things under your feet! Groping around, you discover what feels like a doorknob on one wall.
Obvious exits: out to the LEGO/Logo Lab

When I typed "go out," I would see

The LEGO/Logo Lab
The LEGO/Logo Lab is a happy jumble of little and big computers, papers, coffee cups, and stray pieces of LEGO.
Obvious exits: closet to the LEGO Closet, center to Center of Centers, library to Library Foyer, and atrium to Third Floor Atrium Landing
You see a newspaper, a Warhol print, Sun SPARCstation IPC, Projects chalkboard, Sign about stuff in the garden, MediaMOO Map, Research Directory, STS Sign, Constructionist Flag, and Train Transfer here.

I soon discovered that if I continued to type commands to go out different exits, I would be exploring a textual simulation of the MIT Media Lab in Cambridge, Massachusetts. To help me explore, a schematic of the building would scroll up my screen when I typed the command *map*. Aha, I thought, here are the hallways and offices where I can hang out and meet others interested in media and the net. The only trouble was that these halls and offices were usually empty. Another command, *@who*, would show me that a number of characters were indeed connected and active in the MOO, but they were all located in rooms—some exotically named—that did not appear on the map.

During those first few times I connected, I would occasionally find another guest in the closet or the LEGO/Logo Lab. It was in the L/L Lab that I took part in my first MOO conversation:

Striped_Guest has arrived
Striped_Guest says "Hi"
You say "Howdy"
Striped_Guest says "Ever been here before?"
You say "Yesterday. for the first time."
You say "How about you?"
Striped_Guest says "Where are you from?"
Striped_Guest says "I come here every day"
You say "Northern New York. How about you?"
Striped_Guest says "Ottawa. Canada"
You say "Oh yeah? My wife and I go up to Ottawa every winter to skate the canal."
Striped_Guest says "The canal sucks"
Striped_Guest disappears for parts unknown.

I remember thinking, why did he—if he was a he—say that? I have been to a number of professional conferences, and no one ever said that something I liked sucked—at least not to my face. Maybe he reacted to my banal middle-class image of wife and hubby skating with all the other banal middle-class couples out on the banal, bourgeois canal. Eventually, though, it dawned on me. He had not made that comment to my face. He was far more removed than that. He typed it to something labeled Guest.

Putting my thin virtual skin aside, I had some questions. Where had he gone? Where was everyone else? Could I go there, too? Shortly thereafter I found the answers to these questions: members of the MOO can "build" rooms that they then own, and when they connect to the MOO, they first appear in their own rooms. Further, anyone can "teleport" to any room that the owner has not locked simply by typing the command *@go <room name>*. So, with a few exceptions, the public space of the MOO is primarily a collection of private rooms.

But this capability for characters to add on to the MOO makes for a complex textual construction. Each person who builds adds more nodes to the database, deepening it and complicating it. If members collaborate, they can create new centers of activity that extend the MOO in unanticipated directions. And the phenomenon of telepresence begins to develop in interesting ways. Indeed, when three media researchers at MediaMOO—one from Australia, one from California, and one from Germany—co-located their virtual offices and synchronized their MOO time so that they could "talk" together, they anticipated the broadband telepresence of Pruitt and Barrett's Corporate Virtual Workspaces.[4] Recall my description of Johann in Berlin and Austin in California standing together in fully immersible VR. It didn't take long for me to sample the poor person's equivalent at MediaMOO.

My first gee-whiz experience occurred shortly after I had become a member of Media, built my own room, named it, and created the descriptions that characters would see when they entered it. I realized I still faced one significant technical problem: I had no way to separate the messages I received from the ones I sent. If I wanted to say something while others were also typing messages, the continually incoming signals would scroll up my screen and obliterate the sentences I was trying to construct. One morning (eastern standard time) I "paged" a character named Wu—that is, I sent a message from my room to a character in another room—asking him how to handle this problem. He paged me back and said I needed a client, a program that would run on my host system and provide an interface for me by, among other things, splitting the screen so that incoming text would scroll above the split while I composed text below the split. Back and forth we typed as Wu explained this to me. The whole conversation lasted about five minutes, he in his room, I in mine. No, that's not entirely correct. Wu was telepresent in his room and sdf in his. Rikki (his IRL—in real life—name) was in South America, Steve was in New York State. At one point he told me to hold on for a moment while he retrieved some information and e-mailed it to me. He then accessed a Usenet document about MOO clients that originated from who knows where and sent it to my net account so that when I finished that MOOing session I could read the document and proceed to find a client from somewhere on the net—which I eventually did.

This technology works, I thought afterward. Wu had taught me something and provided me a service in a short period of time in a friendly informal setting over a great distance and at a very low cost, much lower than if we had used telephones and fax machines. He and I worked together for those few minutes as if we really did have offices in the same building. Our collaboration operated within a representation that seemed to be qualitatively different from any other media I had ever used. Eventually, I discovered that this difference was more complex than I had imagined.

As I learned how to negotiate in this new medium, I assumed for a while that I faced two distinct kinds of difficulties: one technical, the other social. The technical seemed to involve how-to-MOO issues while the social centered on my efforts to establish an online persona and to interact with others. But I soon learned that this distinction was false. On the MOO socialization and technical expertise go hand in (typing) hand.

At Media I became socialized by learning a number of sociotechnical things. I learned how to type messages that would appear to others like this:

sdf says "Hi. It's good to see you again."

and how to type messages that would appear to others like this:

sdf thinks we should probably begin the meeting.

I dealt with issues involving page lengths, word wrap, and split screens. I jumped into conversations and began to learn the discourse conventions.

When I arrived in a room or when someone else entered a room I was already in, for example, I learned to type things like:

sdf waves to . . .

People do this to be friendly, but they also wish to make it clear that the physical person represented by their character name is at the keyboard and ready to participate. Likewise, I learned to narrate my MOO experience to expand the complexity of my response to others, to establish some level of creativity and playfulness (even in so-called professional situations), and to better develop interpersonal bonds:

sdf sheepishly admits to reading only one of those books—and that one was the ez one with lots of pictures.

or

Aja says "OK, who wants to volunteer to put that information online for us?"
Aja says "Hmmmm, well don't everybody jump up at once."
sdf nervously clears his throat and seems to be quite interested in his shoes.

All these kinds of discourse appeared to me to be the representation of interpersonal interaction among all those people who were actually seated at their keyboards in distant locations. Yet this appearance deceived me in a subtle but important way. Examine for a moment the following scene, typical of most MOOs, which opens with the name and description of my room.

The Royal Scam
See the glory. . .
You enter a bright, sun-lit room amidst dozens of green and flowering plants hanging from above. Beyond the plants you notice walls lined with books. At the far end of the room in front of a large open window, you see an old oak desk and Queen Anne's Chair.
sdf and Hoops are here
sdf waves
Crimson_Tide arrives out of nowhere
Hoops waves to sdf

sdf waves to Crimson_Tide
Hoops says "Hi, Crimson_Tide"
Hoops hears the tinkling of wind chimes and sees Rose_Darling descending slowly and gracefully into your midst.
sdf waves to Rose_Darling
Crimson_Tide hugs Rose_Darling
Rose_Darling gives Crimson_Tide a big hug
Crimson_Tide says "Hi, Rose"
Hoops waves to Rose_Darling
Crimson_Tide jumps up, clicks his heels together and pops a certs (with a sparkling drop of retsin) into his mouth.
Rose_Darling pokes Crimson_Tide
Hoops bonks Crimson_Tide onna head
sdf says "Ok, looks like we're all here. Shall we begin?"
Crimson_Tide aieees
Crimson_Tide smiles and rubs his forehead
sdf eyes Crimson_Tide warily
Rose_Darling says "I'm ready."

As I gained MOO literacy and became a regular, I conversed in the kind of language you see in this quotation. I began to type statements like "sdf eyes Crimson_Tide warily" because eyeing warily is a phrase commonly used when one wants to narrate generally good-natured doubts about someone. (Quite often characters eye themselves warily, especially after typing something that suddenly appears to them to be pretty stupid. I've done it numerous times.) What I did not realize for quite a while, though, was that while I was typing all my statements, others were not. They were giving a variety of commands that generated these statements. Indeed, nearly every statement in the previous conversation — other than the direct quotations — can be delivered by invoking what are known as feature objects (FOs). The following list includes a few of the many available at MediaMOO.

*ghug <player>*. Sends a quick hug of greeting to the specified player.
*wv <player>*. Delivers a friendly, long-distance wave to <player>.
*botcl <player>*. If someone says or does something that you find uproari-

ously funny, you can let everyone know by bouncing on your chair, laughing.

*grouphug <player> [<player> . . . ].* Grouphug allows you to hug more than one player simultaneously. Just type in grouphug with the player names separated by spaces, and let them know you care!

*jsigh.* Gives a loud, gusty sigh of contentment and relief. A good kind of sigh . . . and a stagey sigh.[5]

Not only can statements be objectified as features, but all representations of persons, rooms, and other textual devices are objects as well. This is the result of the object-oriented programming language through which MOO software is constructed (MOO stands for MUD object-oriented). At MediaMOO sdf is object 7211. My room, the Royal Scam, is 7857.

Given the manipulation of feature objects in MOO talk, what is happening is not the representation of interpersonal discourse but, as John Unsworth has argued, the representation of the representation of interpersonal discourse. Several levels of abstraction separate the person at the keyboard from the persona appearing in the MOO. And the MOO technology shapes these levels in subtle but significant ways. "MOOs in general take shape under twin forces not unlike fate and free will," says Unsworth, "where free will is what we always have understood it to be, but where the role of fate is played by the operating system in which the MOO is embedded. . . . computer operating systems are historically and culturally determined."[6] In the MOO, then, free will means that I can say anything I want. I can create any FO I want and use it any way I want. But fate means that socialization is, in part, the communal immersion within the constraints of the technology. It means that the technology shapes expression (in a small way) as feature objects go from the real to the hyperreal. That is, at first users invoke FOs on the basis of their off-line perceptions (do I want to say that sdf hugs someone? Am I—Steve, sitting in New York with his hands on the keyboard—a hugger?), but over time feature objects can lose their connections to real hugs and become what one "does" when one meets another player in the MOO. So, in this case, as interpersonal relationships become increasingly virtualized, the representation of the representation of affection replaces affection — but convinces us that we experience affection. Finally, fate means that the power to shape discourse belongs, in part, to the programmers (who will become a smaller percentage of MOO participants as more people enter these virtual worlds).

Because I am not nor ever will become a skilled MOO programmer, my ability to manipulate a large but limited set of feature objects created by others shapes my online persona. And that makes for a compelling game of self-construction. I can watch as I create and manipulate my selves within the confines of the virtual world I've entered. But does this game really connect me to others? Or is this primarily a navel-gazing exercise that enables me to look at these abstract selves I've created set against the personas I've encountered scrolling up my screen? A number of proponents have touted the benefits of persona development and interpersonal growth that individuals can achieve through the net. And apparently the net enables two qualities that individuals can find empowering: anonymity and intimacy.

The net is often characterized as the proving ground of the postmodern self: a self that is fragmented, ever changing, ambiguous, and perhaps even liberated. In particular, the text-based net offers the possibility of human relations devoid of certain biasing factors, such as race, age, and gender. Individuals are disembodied and, in theory, unbound by the body's constraints. But the situation is not that simple. First, it is a monumental task to develop close relationships while keeping the particulars of the body anonymous. Second, cultural biases that exist offline can be made manifest online in a variety of ways; therefore, the net is rarely a refuge from those biases.

Those who regularly consume images of the net generated by national news media will have learned about net sex, net marriages, and net friendships, the apparent bonding of individuals who met online. They will also have learned about harassment and flaming, the explosive anger generated by individuals engaged in online discussions. Indeed, the net in general and MOOs in particular are compelling in part because of the emotional intensity of interpersonal relationships that develop online.[7] And as the research analyzed by Martin Lea and Russell Spears shows, intense interpersonal relationships can develop via computer-mediated communication despite its narrow bandwidth—that is, even though no face-to-face or nonverbal communication takes place. Lea and Spears argue, however, that this intensity means that personal differences can also be communicated and felt, generating power dynamics within interpersonal relationships online. These relationships thus are not necessarily egalitarian even if they are anonymous.[8]

Although racial bias has been explored in at least one study, cultural biases online have been most clearly seen through the lens of gender relations.[9] In an overview of opinions about gender on the net, for example, M. H. Dickel offers both pessimistic and optimistic visions of gender relations in cyberspace.[10] One of the negative views is expressed in the opening words of "Gender Issues in Online Communications," a position paper by Hoai-An Truong and the Bay Area Women in Telecommunications: "Despite the fact that computer networking systems obscure physical characteristics, many women find that gender follows them into the on-line community, and sets a tone for their public and private interactions there—to such an extent that some women purposefully choose gender neutral identities, or refrain from expressing their opinions."[11] But can anyone have extensive interactions on the net while maintaining an anonymous, gender-neutral identity? This may be attainable, argues Dickel, through the representations of self enabled by networked virtual realities like MOOs. Indeed, MOOs have been the site for a variety of gender-bending experiments.[12] In the process of reexamining a celebrated case of "virtual rape" that occurred on LambdaMOO, for example, Josh Quittner describes a woman whose MOO character, Legba, takes on a variety of forms and genders at different times. In the rape incident, a character by the name of Mr_Bungle had invoked a feature object that made it appear to others connected to the same room as Legba that Legba and another character were performing graphic sex acts. But Quittner shows us that Legba had masked her gender state at the time of the incident: " 'I like subverting people's ideas about gender and race. During the Bungle incident, in fact, I was not female but a hermaphrodite,' " she says. " 'It was not specifically an attack on a female MOOer.' " [13] Whether this incident constituted sexual assault or was more akin to sexual harassment has been debated vigorously in a variety of forums. Either way it certainly indicates that gender anonymity is no guarantee against online abuse.

Even so, anonymity constitutes some element of control. At Media-MOO the issue of anonymity has a long history of controversy. Most MOOs that emphasize role playing offer their members the chance to remain anonymous—they can decide whether to make their names and e-mail addresses accessible to other members. When MediaMOO began, it too had this policy. But Media, unlike other MOOs, was to be an experiment in professional MOOing. Members of a discipline are rarely rendered anonymous in their professional interactions. At some point, therefore, the "janitors" (these are the individuals, called wizards in most MOOs, who have access to and control over the MOO database and software) changed the policy. From that point on, new characters would not be granted anonymity. (Anyone who wants to find out my offline name simply types "@whois sdf," and my name and e-mail address are revealed.) The policy for all those who had already joined as anonymous characters was to grandfather their anonymity while encouraging them to give up this privilege. Some did, but not all.

The new policy generated heated debate. On one hand, those who thought anonymity was essential to the new kind of virtual egalitarian world disliked the move toward identification. They noted that for most characters the new policy revealed gender and contact information (e-mail address). Gender, they asserted, can have profound effects on the nature of interaction on the MOO; contact information is a potential gateway to further invasions of privacy. On the other hand, those who were identified felt as though there were suddenly two classes of characters, those who could wield the power of anonymity and those who could not. They argued that members of a professional community should not be anonymous to each other. I agreed in general with the second position, although I did not agree that MediaMoo would become a professional community merely by changing the policy. But the debate raised an important question in my mind: Can the interactions of the representations of representations of anonymous characters constitute a public space in which a community can deliberate its common good?

MediaMoo has been the site of a range of public events. Early on, a Cyberspace and the Humanities Conference was organized, and several large-scale MediaMoo parties have marked the inauguration of the moo and subsequent anniversaries.[14] In occasional poetry readings members have taken turns typing their own or others' poetry. In several moo "film" showings text-based moovies were "projected" to all characters in a room. In addition, several attempts have been made to organize regular public gatherings for informal chat. The longest-running and, arguably, most successful public enterprise at MediaMoo has been the weekly Netoric meetings in the Tuesday Cafe.[15]

Every Tuesday at 8 P.M. eastern time, a group of college writing teachers who are dedicated to the integration of computers and of computer-networked classrooms into the teaching of writing meet to discuss a topic of professional interest. This enterprise has been sustainable because it is a means to carry on the business of a group of people who meet at national and regional conferences each year, who read and contribute to a print journal (*Computers and Composition*), and who have personal and professional connections that precede or, at least, supersede their brief weekly moomeets. Some of the regulars connect to MediaMoo as guests each week because it is Netoric and *Computers and Composition* that bind them, not MediaMoo. I have watched MediaMooers who do not read

*Computers and Composition* leave Netoric meetings because the discussion had little to do with MediaMOO.

Yet many experienced MOOers would argue that the technology is not well suited to large-scale events (in MOO terms that may be, say, ten characters or more; the Cyberspace and the Humanities Conference involved more than forty people).[16] Whereas participants in spoken conversations can provide nonverbal cues indicating that they are listening, those engaged in MOO conversations must continually type statements like "sdf nods" or "sdf agrees" so others will know that the typist generating that character is still present at the keyboard. If you are in a MOO room with fifteen other characters, all of whom are actively participating, you will be challenged to follow the conversation. It's like standing in a crowded bar where every utterance is spoken directly in your ear.

Although this factor may contribute to the scarcity of public events at MediaMOO, it doesn't explain why Media has had difficulty in establishing public space for informal, unplanned, regular public interaction. All the events I have mentioned involve some planning and (although the talk during the events is typically loose and playful) some structural formality: a public meeting is arranged for *x* purpose to begin at *y* time; participants should be prepared to discuss *z* topic.

After I had been a member of MediaMOO for some time, I noticed that fewer and fewer characters seemed to be connecting to the MOO, except during the poetry readings and Netoric meetings. When I would connect at random times—either during the day or at night—and type "@who," I'd see only a few characters connected, with usually only one character per room, occasionally two, and rarely more than that. MediaMOO began to seem like a hallway of offices with all the doors closed and only a few people present. Then, in response to a message posted to one of the internal MOO mailing lists, a few members got the idea that the geography of the MOO needed to be changed. What Media needed was publicly accessible spaces to which anyone—guests, newbies, and regulars—

could connect. After much discussion among a small group of veteran MediaMOOers, several enterprising members crafted and programmed a virtual outdoor commons, named Curtis Commons in honor of MOO pioneer Pavel Curtis. This new space was carefully designed so that it connected to the virtual Media Lab and other elements in the MOO geography. The hope was that with this new public construction Media would become again like that third place that had been sought since its inception.

During the process of developing the new place, it was noted that another publicly accessible room had been designed some months before for the same purpose. This virtual grille had attracted participants for only a few weeks. Someone else pointed out that when MediaMOO began, public space had been available within the original Media Lab simulation and had been successful for a while. But Curtis Commons would not become as deserted as these spaces had, would it? Well yes, it would. After much initial genuine goodwill and excitement, very little happened. Now, only the occasional guest appears to be exploring Curtis Commons.

The level of activity can always change, of course. A group of new characters is now adding an interconnected series of places to Media, and this construction has brought some activity to the MOO. But of more than a thousand individuals who have characters on the MOO, based on my admittedly limited and largely casual observations, only a tiny percentage use the MOO regularly, and a significant number of those are new arrivals.

Overall, my experience at MediaMOO tells me that verbal actions on the MOO have social consequences and that there are social enterprises through which individual MOOers connect with each other. Some individuals surely have developed strong and potentially lasting electronic relationships with others via MediaMOO. But overall, MediaMOO has not succeeded as a third place. It has not succeeded as a public space where unplanned, frequent meetings among regulars foster the building of a community. I doubt that any MOO can achieve the status of a "great good place" for two reasons. The first, involving privacy and private property, is a problem that can be handled within the technology. The second, involving the context within which third places are found, cannot.

*Privacy and Private Property.* In an environment like MediaMOO it is easy to develop a private existence because the modern capitalist ideology underlying the MOO encourages the private ownership and control of space. The first thing new characters do is to construct their own space. That space operates as a refuge for and a representation of their selves. These private spaces can be locked, and characters can operate from the spaces and between them through the devices of paging and teleportation. In such an environment, one's capital is completely bound by a right of privacy.

The technology of MOOs, however, does not require such a socioeconomic structure, as shown by the Pt MOOt experiment undertaken in 1993 and 1994 in the ACTlab of the University of Texas at Austin.[17] Pt MOOt, a simulation of a small Texas town, was designed, according to its creators, on a "reality model" of social interaction. The project attempted to model community life more closely than other MOOs by making the metaphors of characters' movement more like geophysical movement. Characters could not teleport but had to walk from place to place (a user would type, for example, "go west") and would thereby be forced to occupy publicly accessible spaces. Feature objects that simulated cars and other public transportation were developed to facilitate movement through the virtual landscape.

In addition, the economic system of the MOO was based on "quota," which was essentially the number of objects a programmer could create within the MOO. One could earn quota by undertaking any one of the jobs established by the city planners (Pt MOOt's version of janitors or wizards), and economic systems could evolve from this source of capital. A group of individuals could share quota, for example, so that objects could be built for the group and owned by the group, or individuals could choose to try to amass quota and objects for themselves. A system of government evolved to resolve complaints. For those who chose not to work, there was welfare: a minimal extension of quota. Other simulated facets of community life were in development, such as a variety of work and entertainment centers, and the inevitabilities of weather, crime, sickness, and death.

Pt MOOt did not succeed for a number of reasons (it closed in July 1994), but its creators recognized that a community is a complex phenomenon involving a vast array of social interactions. And although I do not believe that that complexity can be simulated in real time, I agree that it is only within that array that community bonds develop.

*An Impoverished Bond.* The concept of a MOO as a virtual world that can embody a community is tenuous. A third place cannot exist separate from a locality because it exists only in comparison to its neighborhoods, to local work, play, and family life, to the institutions and formal rituals that encompass daily life. As Oldenburg says, "Third places thrive best in locales where community life is casual, where walking takes people to more destinations than does the automobile, and where the interesting diversity of the neighborhood reduces one's reliance on television. In these habitats, the street is an extension of the home. Attachment to the area and the sense of place that it imparts expand with the individual's walking familiarity with it. In such locales, parents and their children range freely. The streets are not only safe, they invite human connection."[18] Third places are not representations of representations of social interaction. Third places are not about the abstract camaraderie of a MOO.

In MOOs people complain about time zones. The Australian, Californian, and German have difficulty finding the best time to meet. In the great good places, everyone is there because they share the time: it's after work; it's a Sunday afternoon. They share the cold, the heat, the local economy. Third places are lived with all our senses. If I had to pick a physical analogy for a MOO, I would say that it most closely resembles an airport bar: it may have bartenders, it may serve drinks, it may even have a brass rail and a piano, but most connections among its clientele are fleeting and its purpose is primarily to offer momentary gratification to transient individuals.

Media MOO and other enterprises like it are intriguing experiments generated by innovative people of goodwill. These enterprises and their offspring (such as broadband audio and video MOOs) will be useful and important for a variety of purposes. But to infuse them with the capability of becoming virtual worlds is to misplace their function dangerously.

## 5 Seeking Public Space on the Internet

The hopeful say that even if so-called virtual worlds risk becoming solipsistic time sinks, certainly other features of a worldwide electronic communication network like the Internet offer a new kind of space for public deliberations of the common good. The net seemingly enhances social empowerment. In early 1990 Lotus Development Corporation announced a product called Lotus Market-Place, a database of information about 7 million U.S. businesses and 120 million American consumers. The product was the result of a joint effort between Lotus and Equifax Credit Corporation; Lotus developed the software and Equifax provided the demographic information. A purchaser could use the database to target businesses and individual consumers for such marketing efforts as direct mail campaigns. In January 1991, less than one year after it was announced but after a public outcry and protest conducted through the Internet, Lotus dropped the MarketPlace project. Langdon Winner called this a "firestorm of computer populism" and referred to the opposition between the e-mail-generated protest and the Lotus product as the "battle of Liberating Network versus Menacing Database."[1]

In May 1994 the Digital Freedom Net began to publicize the banned writings of political dissidents by making them available via the Internet. The initial works placed on, or posted to, the net came from such people as Chinese dissident Wei Jingsheng and Indonesian novelist and dissident Pramoedya Ananta Toer.[2] Another of the many examples of online international political and social activism is PeaceNet, a nonprofit service in San Francisco, through which groups like Africa Watch and the Lawyers Committee for Tibet publicize abuses of political and human rights.[3] The PeaceNet gopher provides information about human rights violations occurring anywhere from Kurdistan to Ethiopia to the United States. Likewise, through the 187Resist gopher, created by a group organized to fight California's Proposition 187 immigration referendum, one can access articles, official documents, and research concerning Proposition 187. One can also subscribe to a 187Resist e-mail list. Through the Yahoo home page on the World Wide Web one can connect to the home pages of dozens of social action sites, such as the Electronic Democracy Forum, the Electronic Frontier Foundation Online Activism Organization List, the Political Participation Project, the Activist's Oasis, and the Women's Leadership Network. Through Act Locally! one can connect to Ozone Action, Electronic Frontiers New Hampshire, EcoNet, Earth Watch, Pure Water Alliance, and Save the Bay. The net can penetrate national borders, giving voice to lone dissidents, oppressed minorities, and citizens' sociopolitical movements.

Further, the net levels hierarchies. Through the interest group devoted to religion on the commercial service Prodigy, for example, as many as thirty thousand people participate in online religious discussions without the hierarchies of traditional churches. "Among those members are priests, rabbis, and lay people," reports David Gonzalez, "all of whom meet on equal footing, which [the moderator of the discussion group, Lynne Bundesen] sees as one of the service's greatest assets. 'I think computer on-line services are democratizing religion,' she said. 'Each person has as much say as another person.' " That belief echoes the views of those who see the net as leveling political playing fields. Candidates for office, political parties, and special-interest movements are going online in large numbers, enabling more voices to be broadcast to more people than ever before. In an article about online debates among the candidates for governor of Minnesota, one of the candidates noted that "on the Internet, it doesn't matter whether Lincoln is taller than Douglas, . . . you can only judge our words and our ideas." The net, argues James Gleick, is the ultimate grassroots phenomenon: "Here's the Internet, a world controlled by no one, like a vast television station without programmers or a newspaper without editors—or rather, with millions of programmers and editors. It's a frontier, befitting its origins: unruly, impolite and anarchic. But also democratic." Michael Strangelove asserts, "From this point onward, everyone wired to the Internet owns a printing press. . . . The means of mass communication has been democratized."[4]

Power to the connected. Through the net, the masses apparently wrest control from elite opinion makers and entrenched policy makers. The vision is that the net becomes so complex and decentralized that no power structure can control it. It is the means by which revolutions are carried out. But one person's revolutionary is another's terrorist. If the net enabled paramilitary hate groups to organize and recruit, would everyone call it liberating? Does it provide the means by which communities can all deliberate about the common good, or does it increase the babble of special interests? Can something be both anarchic and democratic, as Gleick claims? In light of such questions, I shall examine more closely the case of Lotus MarketPlace.

In her powerful analysis of the Lotus affair, Laura Gurak shows us that this event—a public protest which forced Lotus to cancel a product that appeared to be a threat to individual privacy—was far more complex than just a letter-writing campaign.[5] Yes, the protest did involve the circulation of thousands of e-mail messages alerting people to the apparent dangers inherent in the Lotus product. And yes, a significant number of those messages were sent directly to the personal e-mail address of the president of Lotus Corporation, jamming his e-mail inbox. But this protest was not simply a case of citizens rebelling against an oppressive corporate action. Gurak calls it "a broad-based social action" via the net, but I believe that her own analysis undermines that claim. The case was really a power struggle among factions within and surrounding the computer industry.

According to Gurak's meticulous reconstruction, more than six months passed between Lotus's announcement of the product and the first online protest. But during those six months, members of the lobbying group Computer Professionals for Social Responsibility (CPSR, an organization to which I belong) began a campaign to target Lotus MarketPlace as a threat to the privacy of individuals. This anti-Lotus crusade led to a November 1990 *Wall Street Journal* article explaining the concerns being raised about the product. According to Gurak, the article "was in part a result of the advocacy efforts of CPSR [and] was picked up and widely circulated on the computer networks."[6] It was this event that spurred the online protest.

What is most important to learn about the protest is that the key participants did not engage in reasoned deliberation about the product and its implications for the right to privacy. Instead, they fanned emotional flames: "The cyber-forum provided an open-ended place for debate and dialog where all participants had the potential for an equal voice. Yet along with this democratizing potential came some problems for deliberative discourse. On the other side of this powerful forum were the highly emotive, sarcastic, and sometimes inaccurate messages that circulated, thus holding the community together through their unique ethos but also creating what at times seemed like an 'electronic mob' quality to the protest. . . . While the *protest* itself was successful, the *debate* about the product was not quite as much of a success."[7] In addition, both sides were not participating equally in the debate. While the "electronic mob" was spreading the word about the product, Lotus was providing little online response. In her analysis, Gurak shows how Lotus management did not understand the ethos of the Internet and was unable to use the net effectively in the company's defense.

A close look at this event thus shows it to be not a network of social action but a narrower effort spurred by Washington lobbyists and, to a lesser extent, the *Wall Street Journal*—hardly a bastion of populist liberation—which then spread quickly through the net to a large number of special-interest techno-elites in and around the computer industry. The net became a tool in a power play against a foe that did not know how to use the new technology. (The glory days when the big institutions and corporations did not have a powerful presence on the net are long gone. If the Internet was ever just a citizens' net, it isn't any longer.) Is the case of Lotus MarketPlace an example of the net's democratizing potential? The net may give a printing press to everyone connected, but it does not necessarily increase the quality of deliberative rhetoric for the common good.

In a keynote address to a CPSR conference, Benjamin Barber analyzed the democratizing potential of networked communication technologies.[8] In doing so, Barber discussed the qualities necessary for revitalizing modern democracy and related each to the capabilities of the net. Three of the most important qualities, noted Barber, are decentralization, participation, and deliberation.

*Decentralization* appears to occur on the net, Barber said. Indeed, the great promise of the net should be the development of entities like viable, lively neighborhood networks. But the converse of local nets is the decentralized access to global nets in the form of multiple channels or spectra. The impulse of the marketplace is toward the global and the centralized, and the technology of the net might merely offer a vast array of access points to global products. Decentralization simply would refer to a scenario in which many channels competed with each other to deliver similar products on a global scale. Most of the important civic issues would be either decided away from public deliberation or impervious to public deliberation.

*Participation* can be enhanced by the net, Barber noted, depending on the issues that engage individuals. One of the dreams of universal access is that everyone will be able to participate in the democracy in ways that would not be possible without the net. But national and global special-interest forums on the net operate like global talk radio, only participants don't have to speed-redial for an hour to get through. In this cacophony, issues are cheap, opinions are cheap, and strong reasons and evidence are hidden or compromised by the sheer amount of discussion. The net is so vast and omni-vocal that the power of any one voice is lost and most of what is discussed never goes beyond the net.

*Deliberation* is not enhanced by the new communication technologies. Democracy, noted Barber, is a process "steeped in slowness." The net, however, proceeds at unprecedented speed. Gurak pointed out the effects of this speed on the nature of the Lotus protest: "Speed of delivery was an important factor. Notes traveled quickly back and forth between personal email accounts, conferences and newsgroups, company bulletin boards, and other online forums. Participants could easily cut, paste, and append information to and from these notes, and many protesters were aware of this speed and simultaneity in their use of phrases like 'pass it on.' "[9] This process resulted in messages that were increasingly anonymous and unaccountable.

The increased use of graphic images in the construction of arguments on the net also inhibits deliberation. Democracy, said Barber, is "rooted in words," and new communication technologies are enabling images to sway what passes for public debate in ever more pervasive ways. The impact of television images on political campaigns is well known, and now we stand on the brink of broadband, interactive networks that will give every special interest not just a printing press but a television station. The World Wide Web, for example, is primarily a graphics delivery system, a presentation medium masked as an interactive network. Deliberative rhetoric is defeated every time by image, and the net will soon be the image deliverer par excellence.

So the process by which we analyze and argue the best course of action is continually weakened in the electronic net. It is afflicted by two powerful forces: (1) carefully crafted images, which persuade regardless of the substance, and (2) the overwhelming number of arguments generated on the net and the speed at which these arguments rise and fall. As a result of the dearth of meaningful public space and the impotence of public rhetoric, engagement in public life is declining and the solitary but globalized individual is becoming more prevalent.

At the center of this globalized individual is a devastating paradox: the perfect consumer is one who believes that he or she has individual tastes and individual rights but whose desires can be satisfied by mass-marketed products. The fundamental challenge of the producer is to encourage consumers to be concerned with the self, and simultaneously to fulfill the needs of many individuals. For this task the net is the perfect machine. Since the inception of mass media, in particular the saturation of television, we have seen the rise of a consumerism whose purpose is to identify and celebrate the self. As Leslie Savan illustrates, this irony is the core of advertising:

> When Monty Python's mistaken messiah in *The Life of Brian* exhorts the crowd of devotees to "Don't follow me! Don't follow anyone! Think for yourselves! . . . You are all individuals!" they reply in unison, "We are all individuals!" That is advertising in a nutshell.
> Advertising's most basic paradox is to say: Join us and become unique.[10]

The reach of this contradiction is increasingly transcending cultures, as Teresa J. Domzal and Jerome B. Kernan note in their analysis of the potential of global advertising.[11] The most powerful communication technologies are enabling producers and advertisers to find a global market for the expression of the self. The ads and images that reach this market combine universals with specifics—localized argot that makes the transcendent seem culture-bound.

The primary market for global advertising is the "postmoderns," as

Domzal and Kernan call them, the consumers born after World War II who inhabit all post-industrial nations and many other newly industrialized nations: "These people have known a world of expanding production and consumption, inflation, and interdependent economies. Their globe has been telescoped by television and homogenized by popular entertainment; political struggles or box-office hits anywhere are struggles or hits everywhere." Recall Douglas Rushkoff's argument that it is the young — these postmoderns — who have a symbiotic relationship with the net, who see the net not as a means but as an other. This other, unfortunately, is not engendering the radical counterculture that Rushkoff describes; it is delivering individuals to mass advertisers. And this delivery begins, according to Domzal and Kernan, with products that most easily express the globalized self: fashion and food, the substances one puts on the outside and the inside of the self. "In such categories as fashion and food, a global 'beachhead' clearly has been established and the trend is inevitably in that direction. Accordingly, we can expect an increasing incidence of global markets driving the economies and cultures of individual countries, particularly Western ones." [12]

Domzal and Kernan go on to make the obvious point that advertisers have a "clear opportunity" to reach the postmodern segment of the globe: "The emergence of what amounts to a world capitalism, coupled with a concomitant rise in 'media imperialism' (Schiller 1989) — films, magazines, popular music, and radio and TV programs are essentially equivalent around the world — give advertisers access to both the products these segments want and the communications technology necessary to reach them." [13] The net, with its interactive tendrils wending their way into every home, is the means by which advertisers can fine-tune desire and gratification more efficiently than mass-communication technologies allow. What we will see is a ballet of intelligent agents and interactive advertising.

Intelligent agents are software programs that will personalize each individual's use of the net. Even now, the Internet alone is more complex than any one person could ever imagine. As its complexity increases by orders of magnitude and as other elements of the net, like the expanding numbers of broadband channels (for example, cable television), continue to proliferate, individuals will employ search systems to find items of personal interest throughout cyberspace. As these systems become more sophisticated, they will change the nature of the individual's relation to the net. As described by Pattie Maes, a noted researcher in the field, intelligent agents become Rushkoff's cyborgian other: "Instead of user-initiated interaction via commands and/or direct manipulation, the user is engaged in a cooperative process in which human and computer agents both initiate communication, monitor events and perform tasks. The metaphor used is that of a *personal assistant* who is *collaborating with the user* in the same work environment. The assistant becomes gradually more effective as it learns the user's interests, habits, and preferences (as well as those of his or her community)."[14] Thus the dumb box or dumb terminal becomes the smart network, a device that participates in the construction of one's identity. The statement "I am what I access" becomes "I am what is accessed by my agent."

"Wait a minute," you might say. "I will program my agent to exclude advertisements." So would I, of course. But efforts are under way to develop interactive advertising in which the mere process of using the net involves us in commerce. "We're going to find ways to make the viewer want to participate with the advertising message," says Michael Drexler, president of the media group at the New York ad agency Bozell, Jacobs, Kenyon and Eckhardt. As Savan shows us, this is not a new idea: "In 1953, a cartoon show called *Winky Dink and You* became the first instance of interactive TV, albeit in a low-tech kind of way. *Dink* host Jack Barry told kids to put a special piece of cellophane—which they could get along with a special crayon by ordering the Winky Dink Kit—right onto the television screen. Then the kids would trace and connect special dots off the monitor—and presto!—they'd receive a secret message!"[15]

So far, most efforts at interactive advertising are not quite so compelling (the August 1994 installation of so-called interactive advertising by McDonald's on America Online involved clicking on items like "What's New at McDonalds" and "McHistory"),[16] but more powerful possibilities are beginning to take shape. G. Pascal Zachary lists potential scenarios in various stages of development, including "The Virtual Peddler," in which your intelligent agent is endowed with the persona of a celebrity who suggests products and services that might interest you. "Imagine Candice Bergen asking 'Have you seen the new power tools from Black & Decker and Sprint's latest calling rates?' and then going on to say, 'Dinner is at 7 and you have six pieces of e-mail unread.'" Or how about the "I Spy" scenario, in which your every keystroke or input message is collected by intelligent agents to build a portrait of you as consumer. "Such use of data culled from interactive consumers might alarm privacy advocates, but it will give advertisers an unprecedented ability to mold their messages to individual tastes." And my favorite scenario is "The Never-Ending Commercial": as you respond to an ad, you be will prompted with another related possibility. Once the the ad begins, it " 'will evolve on the fly in response to viewer preferences,' says Jeff Apple, chief executive officer of daVinci Time & Space, which is creating interactive TV programs for children that will include interactive advertising. Watching a car commercial, a viewer could ask to look under the hood, take a simulated test drive, then request details on performance. 'If a viewer is responding, the ad can keep on going—almost like a real-life sales pitch,' Mr. Apple says."[17]

So from childhood onward we will be interacting with advertisements, building intelligent agents or cyborgs, empowering ourselves in the decentralized economy, making well-informed choices in our democracy of products and services, fine-tuning our individuality, and gaining membership in the global village. But when we virtualize human relations, notes David Ehrenfeld, we are no longer in touch with the essential ingredients of community, "for at the end of the day when you in Vermont and your E-mail correspondent in western Texas go to sleep, your climates will still be different, your soils will still be different, your landscapes will still be different, your local environmental problems will still be different, and—most importantly—your neighbors will still be different, and while you have been creating the global community with each other, you will have been neglecting them." [18]

Believers in virtual community might say that I am blind to new kinds of relationships—healthy, life-affirming virtual relationships—enabled by the technology; that I am devaluing the therapeutic potential of online personal development, especially in light of the difficulties that face anyone who is different from the dominant image of a community and culture. The net, they might point out, houses discussion groups and information archives for cancer victims, for AIDS victims, for recovering alcoholics, for the victims of domestic abuse, for the victims of sexual harassment, for religious minorities, for gays and lesbians, and for thousands of others who may need support that is not forthcoming in their offline lives.

I do not doubt that virtual experimentations with the self and with the relations of that self with others can be liberating. But I can't help feeling that the situations that call for these benefits reflect deficiencies in our geophysical communities. The institutions, the families, and the social relations of our offline lives are unable to include and celebrate those who are different, to care for and heal those who are hurting. If the net becomes the only recourse, then our geophysical communities are lost.

# Part Three

## The Nomadic

## Individual

Most of what needs to be done by way of human intervention will be done out of the home. One forms an eerie vision of the high industrial future: a vista of glass towers standing empty in depopulated business districts where only machines are on the job networking with other machines.

— *Theodore Roszak,* The Cult of Information

## 6 Telecommuting

There's no better indicator of the seductiveness of an idea than its presence in an uncritical television commercial. Picture a Bell Atlantic advertisement I saw recently in which a group of harried office workers gather for a meeting. The actors playing the office workers appear unhappy, uneasy, tired, and slightly fearful. Then the boss tells them about a new company policy: they can all work at home. *At home!* The words lift the darkness from the workers' faces. The actors emote happiness and the beginnings of excitement.

Another example is the Compaq computer advertisement that tells us that Mozart composed during the wee hours of the night, Hemingway wrote in the morning, and Edison was inspired by the darkness. Then it shows us a woman walking past her sleeping children as she enters her home office. The narrator says, "At Compaq, we understand that ideas aren't governed by a clock."[1]

How about examining the words of education historian Diane Ravitch as she unintentionally creates an advertising image in her predictions of education being transformed by the net: "In the not very distant future, the technology will exist to teach anyone anything at anytime. . . . In this new world of pedagogical plenty, children and adults will be able to dial up a program on their home television to learn whatever they want to know, at their own convenience. If little Eva cannot sleep, she can learn algebra instead. . . . Young John may decide that he wants to learn the history of modern Japan, which he can do by dialling up the greatest authorities and teachers on the subject."[2] Ravitch's projection reads like a treatment for a thirty-second spot. (Who knows, maybe one of the creative people at the advertising agency retained by AT&T had read Ravitch before designing its infamous "Have You Ever . . . ? You Will" ads: "Have you ever answered your teacher's question . . . from your bed? You will.")

One of the most powerful seductions of the net is that it will take us back home. It will enable us to telecommute to work and will create ultra-accessible virtual schools. It will allow us to spend more time with our families and stay closer to our communities. Given the dangers of the world—crime-ridden cities, terrorism, suburban alienation and decay— the vision of the home as a safe house connected to the world is compelling. Further, the changing economy enables and encourages all forms of telecommuting. As work becomes increasingly knowledge based, workers become knowledge manipulators. Work involving knowledge—whether on the job or in school—is fertile ground to be digitized and enacted via the net. Telecommuting to work and school represents the great practical promise of the net and justifies its reach.

And telecommuting completes the movement of social activities from public places to private ones. Radio and television brought home the entertainment that at first simulated the diversions of theater, dance hall, pub, and nightclub, then either obliterated or transformed them. Every other electronic advancement, from the video cassette recorder to interactive networked video games (to a future filled with "killer apps" like VR sex simulators), augments the capabilities of these diversions. We can play all we want at home, and now telecommuting seems to deliver our work and our school there, too.

Peg Fagen, for example, lives in Massachusetts and works for a company in New York. She is an environmental engineer, and in her home office she has all the necessary items to do her work: dedicated phone and fax line, computer and modem, e-mail and file-transfer connectivity, copier, and so on. Her children, all of preschool ages, are supervised by a visiting sitter while Peg works, largely uninterrupted in her office. Peg is one of the digitally select—those people whose expertise and net connections enable them to do valuable, highly skilled, and challenging work from their homes or mobile offices.

Peg began working at the New York office of the firm in 1989. Less than two years later she and her husband moved to Massachusetts so that he could take a job there. With that move and with the hope of starting a family, Peg asked her employers if she could begin telecommuting. At that time the company was a small, private operation with a flat management structure enabling most employees to work with the owners in a close-knit group; therefore, the company's managers were familiar with the quality of Peg's work and aware of her value to the company. They agreed to try such an arrangement by removing her from their employ and hiring her as an independent contractor. After less than a year both Peg and the firm were happy with her ability to do the work, and she was rehired as a regular employee. Seven years later she is still telecommuting each day.

Peg was the first employee in the firm to telecommute. A number of others have requested a similar arrangement since she began the experiment, but only a few of those requests have been approved. As Peg herself recognizes, much of the work of the firm requires that its employees be at the office dealing with co-workers and clients. Most of her work, however, involves analyzing and reporting on environmental pollution data. She identifies methods of treating contaminated media (such as groundwater or soils) found at clients' sites, and she communicates with companies and laboratories that specialize in remediation methods.

She has disciplined herself not to mix work around the house with her engineering work. When she goes into her office, she is at work. When the sitter leaves or when the children awaken, she is at home. If the day's work is not completed when the sitter leaves, then she returns to her office after the children have been put to bed. In addition to her ability to manage her time productively, one of the keys to her success, she says, is her relationship with one manager at the firm's office. Because she had worked for that person before she began telecommuting, the manager understands the kind of resource that Peg provides and can steer appropriate work her way. Peg recognizes that it can be difficult to develop a solid working relationship long-distance with someone she has never met. Overall, it has been easier for her to work with people she knew before she left the office. Peg also believes that it helps if the manager responsible for choosing the staff for a project knows the potential team members personally. As a result, she visits the office every few months to see both new and old employees.

The firm evaluates Peg from several perspectives. The managers for whom she works offer qualitative assessments of her performances while tabulating her "utilization rate"—her rate of billable hours. The firm would like to see a rate higher than 85 percent; Peg always exceeds this rate. Her one disadvantage as a telecommuter involves her progress within the company's hierarchy. Co-workers with similar education, on-the-job experience, and time with the company have moved up the company's management ladder faster than Peg. She believes that although she can work as a technical analyst from her home, she cannot so easily manage others remotely. While she anticipates the possibility of giving up telecommuting after her children are all in school full-time, Peg is thankful that her arrangement has enabled her to stay current in a highly technical field that evolves rapidly. If she had taken five or six years off to raise children, she would have been out of touch with the science and technology of the field.

Peg is certainly not alone in her method of employment. More and more workers are doing all or most of their work from wherever their electronic connectivity allows. The issue is no longer the feasibility of telecommuting but the quality of life that net-infused work will create.

When an earthquake hits Los Angeles or San Francisco or any other major metropolitan area, the advantage of telecommuting becomes quite visible. Even after the immediate crisis subsides, hundreds of thousands of people who commute in cars, buses, or trains have trouble getting to work. The telecommuters, however, have no difficulty (assuming that their jobs and telecommunication connections are still intact). Thanks to technological developments making networking cheaper, more capable, and more secure, more people were able to switch to telecommuting to cope with the Northridge quake in Los Angeles in 1994 than could switch four years earlier after the Loma Prieta quake near San Francisco. The Los Angeles County Telecommuting Program estimated that before the Northridge quake there were nearly five hundred thousand tele-commuters. Immediately after, that number jumped to seven hundred thousand.[3]

But it doesn't take natural disasters to spur this trend. Telecommuting is growing rapidly. According to Link Resources, a technology research firm, there were 7.6 million telecommuters in the United States in 1994, a 15 percent increase from 1992. There were also 24.3 million self-employed home-based workers and 9.2 million after-hours home-based workers for a total of 41.1 million, or one-third of the adult workforce. Between 1989 and 1993 this number grew at nearly 9 percent per year. By the year 2000, the number of U.S. telecommuters alone is estimated to reach 25 million.[4]

This growth represents the expansion of the technology that enables telecommuting, but it also represents a shift in the definition of workers' productivity. Most jobs now involve manipulating storable and retriev-able knowledge. The results of that manipulation define productivity. Where the work is done, or who sees it done, is of decreasing importance. This shift has powerful implications.

Early indications are that telecommuting can cut employers' costs. Companies can slash overhead by eliminating or reducing the costs of real estate, maintenance, energy, landscaping, security, and some kinds of insurance.[5] In April 1993, for example, Compaq Computer Corpora-tion moved all sales personnel to home or mobile offices. In the process

Compaq reduced the number of salespeople from 359 to 224, cutting its sales cost by 10 percent. Many other companies have similar strategies. By sending large numbers of its employees home, IBM-Midwest cut real estate costs by 55 percent and increased the ratio of employees to workstations from 4 to 1 to 10 to 1. A large New Jersey consulting firm, Gemini Consulting, has no central office for its sixteen hundred employees who are scattered across the country, communicating via the net and being trained by CD-ROM. Perkin-Elmer Corporation, which produces scientific equipment, closed thirty-five branch offices by moving its sales and customer support workers to home and mobile offices.[6]

Evidence also suggests that telecommuters can be more productive than they were in offices. Within the first year of its restructured sales operation, Compaq increased its revenue even though its profit margin continued to drop as a result of competitive price-cutting. With no central offices or regular hours, the sales force can be in contact with customers anytime they are needed.[7] Similarly, telecommuters for branches of the Los Angeles County government work better at home than in the office. Joel Kotkin, of Pepperdine University and the Center for the New West, reported in 1994 that the Los Angeles County assessor's office "found telecommuters process their work 64% faster than office bound employees and that overall productivity rose 34% once telecommuting was instituted."[8] But these are not the only benefits, argues Kotkin. The practice of telecommuting is also having positive social and economic effects.

Using Los Angeles as an example, Kotkin notes that hundreds of thousands of commuters in the region must take trips of an hour or more to work every day. Those commutes will continue to get longer, he maintains, as the number of commuters continues to increase. When emergencies like earthquakes occur, commuting time balloons. Historically, the solution to congested roadways has been to increase capacity by building more and better freeways and faster trains. The better solution, Kotkin says, is to get the people off the road and onto the net. Because 60 percent of the pollution in the Los Angeles basin is from automobile emissions and because the cost of building new roads and train lines far outstrips the cost of developing the telecommunication infrastructure, virtual commuting is the logical response to the needs of both employers and employees. Besides, he argues, telecommuting increases employees' time at home with their families. Telecommuting and home businesses "root workers more deeply in their communities," says Kotkin, "something that could help counter the anomie associated with commuter suburbs."

But in order to be rooted in the community, one must spend a long time integrating one's life into a place. For the digitally select, tele-commuting offers this option. They will be able to pick the place in which they live and stay there if they like, enabling them a new kind of community stability. In the past, external forces like jobs and family rooted people to places whether they liked it or not. Telecommuting affords rootedness by choice. But only the digitally select will be able to work completely online and have freedom to choose where they live. The rest will need to live near their jobs. The problem is that those jobs are becoming increasingly nomadic; the threat of the employer's moving, collapsing, or hibernating makes it difficult for employees to commit to the place where they live. Yet while the net induces instability for em-ployees, note William H. Davidow and Michael S. Malone in *The Virtual Corporation,* it produces flexible employers able to compete in the global marketplace: "While technology will not solve all of the manufacturing problems facing the virtual corporation, it will certainly present it with a host of opportunities. One of these is an unprecedented degree of flexi-bility in which to locate production facilities. With instantaneous world-wide communications it is theoretically as easy to control a factory in Asia as it is to control one right next door. The ease with which production can be integrated around the world will provide companies with greater flexibility than ever in selecting plant locations." [9] If manufacturing is less tied to place in the age of the net, imagine how dislocated knowledge workers may become.

Supporting Peter Drucker's depiction of the emerging knowledge economies, Richard J. Barnet and John Cavanaugh illustrate the trend on a global scale. They point out that the top growth industry in the world is the production, processing, and selling of information and that this industry is infused with utopian visions of the near future. "The prophets of the new age of plenty have promoted dreams of a world where homes are turned into 'electronic cottages' and people interrupt lives of leisure and self-development with occasional work as needed to pay the bills. Thanks to the modem, fax, and fiber optics, postindustrial knowledge workers can plug themselves into the global economy whenever they like and cut out when the spirit moves them. The result will be to revive warm and loving family life that will no longer be held hostage to bumper-to-bumper commutes and other frazzling demand of the economic rat race. Thanks to the huge global market in intangibles, workers will no longer be locked into a place." Such scenarios, say Barnet and Cavanaugh, "romanticize and obfuscate" what is happening. The large bulk of information-era jobs that can be performed in any location—typically jobs that involve data entry—tend to be farmed out to low-wage areas. And increasingly, sophisticated high-tech production jobs are also being moved to cottage industries. Barnet and Cavanaugh cite numerous examples; in 1990, for instance, "Texas Instruments, the Dallas-based semiconductor firm, set up 41,500 computer terminals in its operations across the world, more than one for every two employees, to link up the 'cottages.' To cut costs more and more, microchip design is done in places like Baguio, a mountain city in the Philippines, and in Bagalore, India, where engineering talent costs a fraction of what it costs in Texas. The design specifications are transmitted by computer." [10]

Whom do these vast telecommuting possibilities benefit the most? Corporations have the opportunity to become lean; they can meet the demands of the market more productively than ever before. But these developments increase the disparity between telecommuters and other workers: we see the rise of the digitally select, the digital sweatshops, and the nomadic workforce.

Nomadic, displaced work is not the only dark manifestation of tele-commuting. Even though some of the examples that I surveyed earlier seem to support the rosy glow of the Bell Atlantic advertisement, other evidence suggests that telecommuting—even for the digitally select—does not necessarily overcome separation from home and community. It can, in fact, increase feelings of isolation from fellow workers.

Although telecommuting operates on the premise that communica-tion technologies can handle most, if not all, of an employee's essential communicative acts, the current state of the art leaves some gaps. Ac-cording to a survey of workers at one site by David Spain, an anthro-pologist at the University of Washington, telecommuters are concerned about "losing visibility and career momentum" by being separated from those who would evaluate their performances. These employees fear being unable to prove their worth.[11] Some telecommuters report that mis-understandings increase between workers and supervisors when tasks and responsibilities are not clearly laid out before telecommuting begins.[12] Others discover that they miss the interaction of peers in the workplace. Even though workers can send and receive a massive amount of infor-mation via the net, that information can be a bit thin: "It isn't what you communicate, it's how you communicate," says Lynn Dreyer, human re-sources director for Great Plains Software. "Without some human inter-action it's impossible to build relationships and a sense of trust within an organization."[13] Two of the toughest tasks that face employers, notes Samuel Greengard in his overview of the pros and cons of developing vir-tual offices, are evaluating telecommuters' job performances and generat-ing a constructive corporate culture within companies comprising large numbers of telecommuters.[14] The second task raises a key question: Can even the most sophisticated communication technologies engender a cor-porate culture when most employees are located apart from one another?

Or will people's work lives—essentially separated from work colleagues—become primarily entwined with their personal lives? Separating work from nonwork is a difficult task for the telecommuter because, for many, being connected means being at work.

Karen D. Walker, Compaq's vice president for operating services, notes that at Compaq telecommuters are sending faxes in the middle of the night, for example, and there is concern that employees are working twelve to eighteen hours a day. Says Compaq sales representative Ann Bacon, "You can't leave it behind, because it's always there." [15] Home workers work more hours than office workers and face more pressures to meet deadlines, according to a report by the Daniel Yankelovich Group commissioned by *Mobile Office* magazine. [16] Franklin Becker, director of the International Workplace Studies Program at Cornell University, comments, "The more you get connected, the harder it is to disconnect. At some point, the boundaries between work and personal life blur. Without a good deal of discipline, the situation can create a lot of stress." [17] Telecommuters may not be tethered to a desk in an office complex, but they are tethered nonetheless. The difference is that this tether goes with them anywhere they go anytime they go there. If you are always connected, are you always on the job?

## 7 Default Equals

## Offline

The technologies that currently enable tele-commuting are telephones, fax machines, cellular phones, e-mail, synchronous computer conferencing, and, in some cases, satellite video hookups. All these connections can be controlled by the user. If telecommuters wish to disconnect, all they need to do is log off the computer network, turn off the cellular phone, or turn on the voice-messaging system. These capabilities seemingly fulfill the promise of the techno-topists described by Barnet and Cavanaugh: "Knowledge workers can plug themselves into the global economy whenever they like and cut out when the spirit moves them." Although the ability to disconnect may be beneficial for the mobile worker, it may make employers a bit nervous. Because employees can make a variety of claims about the amount of unmoderated work they do or about the wages and benefits they have earned while unmoderated, Patrick McCarthy, partner in the department of labor and employment law in a New Jersey law firm, suggests that employers follow four guidelines to ensure that both employee and employer are legally protected.

To decrease the possibility that an employee will make claims for wages and benefits that are not due the employee, employers should:

1. Establish and enforce methods for monitoring work done at home.
2. Educate operating managers and supervisors about the necessity for keeping workers within the 40-hour workweek.
3. Use such technology as time clocks, or other equipment that records hours.
4. Give employees explicit written directions not to work more than 40 hours without prior written supervisory approval.[1]

The technologies that enable employers to monitor their mobile workers are being developed. At the 1994 Directions and Implications of Advanced Computing Conference held at the Massachusetts Institute of Technology, one of the pioneers of MOO software, Pavel Curtis of Xerox Palo Alto Research Center (PARC), presented a status report on the next generation of MOOs under development at PARC.[2] These systems involve real-time sound and video so that those connected to the MOO will have available not just synchronous text-based communication but video and audio communication as well. This technology appears to be another step toward the Corporate Virtual Workplaces described by Pruitt and Barrett. As Curtis described the new system, I remember thinking, "Little video cameras watching me watch everyone else. Don't know if I'd like that." To be fair, Curtis placed a high priority on personal privacy and discussed how privacy safeguards are designed into such systems. Nonetheless, I still worry—especially since I learned about another PARC project, Ubicomp.

In an article entitled "PARC Is Back!" Howard Rheingold describes PARC researcher Mark Weiser and his work with ubiquitous computing, or Ubicomp, as Weiser calls it. According to Rheingold's version of Weiser's work, Ubicomp represents the attempt to make computers invisible. This means that the technologies must evolve beyond easy-to-use interfaces and away from virtual reality. Merely improving interfaces makes the obstruction (your computer) nothing more than an easier-to-use obstruction. And VR places your entire working universe within the computer, making the device the ultimate obstruction. In contrast, the goal of Ubicomp is to enable you to work with the aid of computers without ever focusing on working the computers.[3]

Ubicomp, at first, intrigued me. In particular, I was heartened to read one quotation attributed to Weiser as he explained the difference between VR and Ubicomp. VR, said Weiser, "has the goal of fooling the user—of leaving the everyday physical world behind. This is at odds with the goal of better integrating the computer into human activities, since humans are of and in the everyday world" (93). I like a principle that seems to privilege human activity over computer use. At least I liked it until I began to read about some specific technological developments in Ubicomp. One of those developments is what Rheingold described as an active badge: "With an active badge system, every computer you sit down at is your computer, with your custom interface and access to your files, because your active badge sends it information via infrared signals. It is possible to track the locations of other researchers at all times by central monitoring of active badges—a handy tool with Orwellian implications" (94). Orwellian is right. Active badges should scare the daylights out of anyone. As I read on, I expected to be assured that any Orwellian nightmares would be unjustified because of the way the technology is being designed. The following passage, I think, was supposed to placate me:

> PARC is an intellectual playground, full of free spirits. How do they feel about the possibility that Ubicomp might lead directly to a future of safe, efficient, soulless, and merciless universal surveillance?
>
> "Some people refuse to wear badges," Weiser says. "I support their right to dissent. And one principle we go by here is to maintain individual control over who else sees anything about us. . . . The answer will have to be social as well as technical." (94)

I agree with the principle of individual control, and I agree that the answer indeed lies in social controls over technological capabilities. But I am chilled by one particular word: *dissent.* Weiser's use of that word

makes me feel uneasy about the effectiveness of social controls on Ubicomp surveillance outside of organizations like PARC—organizations that
encourage freedom, privacy, and ingenuity among their employees. I can
easily imagine many organizations not so enlightened, not run by people
who believe that privacy and freedom are essential to the organization's
well-being, organizations that would tolerate little, if any, dissent. If we
begin by thinking that refusing to wear the badge is dissent, we are asking
for, as Rheingold put it, "merciless universal surveillance."

I think that Weiser has it backward. We need to begin any consideration of Ubicomp by accepting a simple equation: default equals unconnected or offline. Accordingly, we need to agree on some principles that
can inform Ubicomp policies throughout society. In the final chapter of
*The Virtual Community,* Rheingold quotes Gary Marx on issues of privacy and computing. Marx presents five principles outlined in the Code
of Fair Information developed by the Department of Health, Education,
and Welfare in 1973:

> There must be no personal-data record keeping whose very existence is
> secret.
> There must be a way for a person to prevent information that was ob-
> tained for one purpose from being used or made available for other purposes
> without his consent.
> There must be a way for a person to correct or amend a record of identifi-
> able information about himself.
> Any organization creating, maintaining, using, or disseminating records
> of identifiable personal data must assure the reliability of the data for their
> intended use and must take precautions to prevent misuses of the data.[4]

In light of my concerns about Ubicomp, I want to propose a few principles of my own:

1. The normal state of anyone's computer is off.
2. The normal state of anyone's relationship to computer networks is unconnected.

3. The normal state of knowledge about the location of anyone is unknown—whether that person is connected or unconnected.
4. Connectivity and location are private and must be protected by both
technology and social policy.

I recognize that there are legitimate reasons for a person to be locatable
and even continually observable. The obvious situations come to mind. If
you are in critical condition in a hospital, you want your specialist to be
quickly locatable. If a fire breaks out in your house, you want your local
firefighters to be quickly locatable. But beyond life-threatening situations, many other justifiable reasons may explain why an organization
may need to keep track of its members. And surveillance via computer
badges may be efficient and satisfying to all involved. That is not my
point. When it comes to connectivity, the employer must justify the surveillance. Everyone must assume that only extraordinary conditions merit
surveillance. The requisite argument must not be, "Why do you not want
to wear the badge?"—in other words, "Why do you *dissent?*" The requisite argument must be, "Why do you want me to wear it?" The burden of
proof must be on the watcher, not the watched.

I first made this argument in the October 1994 issue of the World
Wide Web publication *Computer-Mediated Communication Magazine.*[5]
In the April 1995 issue Mark Weiser responded by noting that he and
others involved in Ubicomp research recognize the potential dangers of
sophisticated surveillance devices but also recognize that the technology
can be a source of great good for humanity. Therefore, he proposes his
own principles:

Weiser's Principles of Inventing Socially Dangerous Technology:

1. Build it as safe as you can, and build in all the safeguards to personal
values that you can imagine.
2. Tell the world at large that you are doing something dangerous.[6]

He notes that principle 2 cannot assure that Ubicomp will not be used for evil, but, he says, "I know of no way to provide such a guarantee for any technology. Refusing to work on such technology is the approach of the ostrich. However, I am an optimist. I think that people will eventually figure out how to use technology for their benefit, including if necessary passing laws or establishing social conventions to avoid its worst dangers." Further, Weiser takes me to task for expecting engineering safeguards to make a technology safe and beneficial. (He was responding to the comment I made earlier: "As I read on, I expected to be assured that any Orwellian nightmares would be unjustified because of the way the technology is being designed.") He points out that "one culture's Orwellian nightmare might be another's prayers answered; a technology which could—on the basis of technology alone—prevent one would also prevent the other. I don't believe there can ever be such a technology: a law of physics of morality. Yes, safeguards can be built into any system, such as the checks and balances in a good accounting system. But what keeps them in place is not the technology, but people's commitment to keeping them."

Here I agree wholeheartedly with Weiser. My hope that the good engineers would protect me was naive—which is why I insist that connectivity be in the control of the user. I want to be able to enter into any telecommunication situation knowing that others bear the burden of proof if they want to control my connectivity. In response to this position, Weiser defends the uses of active badges in PARC and celebrates the opportunities that employees have to resist the practice. "It takes courage and intelligence to oppose it when necessary," he says. "It is vital, I believe, to have a tradition of honorable dissent, of supporting those who cry 'No!' when everyone else is swept along." I share his enthusiasm for dissent, but I think we need to force the issue *before* we are told that we can't disconnect.

Is this paranoia? Some may say so. In another response to my original column, David Porush argued that I am advocating an isolationist agenda that is anti-community.[7] He prefaces his position by pointing out that to reject the net is to reject the culture.

> PORUSH's LAW: Participating in the newest communications technologies becomes compulsory if you want to remain part of the culture.

He says that my principles construct a world of discrete isolated individuals. Given the inevitability of network connectivity, he argues, we should try for and hope for the best: a democratic society that will continually seek out generally acceptable ways to protect privacy where it is needed. "Let's not be so hasty to re-define what is the natural or the 'normal' or the default state of our communications and our mutual connectivity. I believe the future negotiations over what information is private and how it will be used will permit mostly satisfactory resolutions to emerge in a democratic society. But assuming that the default state of human communication is private, disconnected, and anonymous strikes at the very core of what it means to be part of a community or a culture in the first place." Here Porush raises an extremely important issue. If we choose not to participate in the means that bind us to communities, we are, in effect, rejecting community life.

I value the interconnectivity of community and family life. I want to be known and findable within my community. Most people who feel similarly connect themselves through sociotechnical systems: they do not choose to have a post office box and an unlisted telephone number. They put themselves on mailing lists and make known their e-mail addresses. Even when they may feel harried or inconvenienced, they allow themselves to be connected because they work at jobs that necessitate connectivity and participate in community functions that necessitate connectivity. Porush argues that most of us are already locatable to some extent. He is right, and I'm glad he is. I also understand his concern about the radical isolationism he sees lurking behind my argument. I reject isolationism. The problem is that I believe there is a qualitative difference between the intermittent and controllable telephone and mail systems and the forced connectivity of Ubicomp. That my e-mail address is knowable does not mean that I am online and definitely doesn't mean I'm wearing an active badge.

Mandatory connectivity is not a question of community; it's a question of power, a question of proof. Who has the burden of proof to establish connectivity? Is it the entity (employer, community, state) who wishes to establish the connectivity, or is it those who would be connected? A community can be torn apart when internal surveillance evolves out of communal cooperation. A community can be oppressed when external surveillance is inflicted on it as a whole. Privacy is not merely a libertarian concern. Privacy is necessary for community development, too.

The greater concern is that universal connectivity engenders not community but a devaluation of placed community. In spite of the seductive images constructed by the advertising agencies of homes all warm, cozy, and connected, the net does not make the home a base from which to communicate with the world. We are beginning to see this in the increasingly peripatetic, net-connected lives of professional men and women.

> The gadgets that let business people fax, phone and surf the Internet from 30,000 feet are from the same technology that once promised to eliminate bothersome travel by heralding a new age of video teleconferencing and the information superhighway—businessmen talking to each other by E-mail and on television screens. Instead, the opposite has happened. Advanced telecommunications have made the office portable, and have propelled more business people into a place called the virtual office—an office-in-a-bag that accompanies them as they work above the clouds. . . . "These people are leading indicators of what's in store for the rest of us," said Paul Saffo, a director of the Institute for the Future, a research foundation in Menlo Park, Calif.[8]

The net does not make the home into the center of our public and private lives but eliminates the center. The result of the omnipresence of the net is that all centers—work centers, school centers, and living centers—become less and less relevant.

## 8 Virtual Schools

*Scenario 1: Amid the Leafy Campus.* Josie Becker is a physics teacher at an expensive private preparatory school in New England. Six days a week he meets classes of eight or ten students in well-equipped labs and comfortable classrooms. Josie and his family eat meals with his students on a regular basis, and he monitors a dormitory a couple of evenings every week. He works hard and gets to know each of his students quite well. At the end of each semester he is ready to collapse and sleep for days.

*Scenario 2: Amid the Flicker of the Screen.* Every Thursday evening last fall I would sit down to the computer at 7:30 P.M. eastern standard time and connect via modem to Diversity University MOO, a multi-user dimension designed for teachers and their students.[1] I was one of three faculty members who organized and participated in a seminar entitled "Rhetoric, Community, and Cyberspace" with fourteen graduate students from rhetoric, communication, and English programs at such schools as Pennsylvania State University, Rensselaer Polytechnic Institute, Carnegie-Mellon University, North Carolina State University, Michigan Technological University, and the University of Minnesota. The participants connected to DU MOO from either their homes, their offices, or computer labs on campus.

Every week one student served as the presenter and two others as respondents. These three would get the conversation rolling by laying out some of the issues raised by that week's assigned reading. The discussions were usually intense, fast paced, and often quite provocative. For an hour or so each of the seventeen of us would be watching the conversation scroll up the screen at a sometimes frantic pace, contributing on the fly, trying to follow conversational threads, agreeing, disagreeing, and challenging each other's interpretations. Some segments of each session would seem nearly chaotic, others incisively focused. The conversation was never dull.

Distance learning is not a new concept, nor is DU the only one of its kind. Through BioMOO scientists have developed text-based virtual labs in which a student may undertake a simulated experiment. In a dissection lab, for example, a student may give commands that activate preprogrammed descriptions of anatomical details an experimenter would see.[2] In some senses, DU and BioMOO are quite primitive examples of tele-schooling; they are text-only systems beset by a host of technical problems that plague the Internet—sometimes systems crash, sometimes they become very slow. These MOOs do not offer the hypertext or graphical capabilities of the World Wide Web, or the broadband capabilities of video satellite hookups. What they do offer, however, is pervasiveness. They are relatively cheap and potentially widely accessible. Neither teachers nor students need to have access to expensive video conferencing centers; enterprises like DU require nothing more than an affordable net connection and telnet capabilities.

Text-only synchronous computer conferencing, however, is primitive not only in comparison to current and future broadband technologies but in comparison to the experiences of the students in Josie's prep school classes. The MOO cannot approach the richness of the interaction in a small class facilitated by a knowledgeable and caring teacher in an environment that encourages and rewards learning. But primitive or not, enterprises like DU represent the beginnings of broad-based public-access distance education, which will expand until most of the students in the world will be educated in virtual schools. No one yet knows if this is a good system or a bad one, but economics will demand it, according to John Tiffin, co-author with Lalita Rajasingham of *In Search of the Virtual Class: Education in an Informal Society.* In his address to the Standing Conference of Presidents of the International Council for Distance Education (ICDE SCOP), a conference of university presidents interested in shaping the future of worldwide education, Tiffin noted that every day billions of students ride cars, buses, bicycles, or trains to school. But this mass transfer of students via the tools of the industrial age is giving way to the mass transfer of students via the tools of the information age. Given the trends in world demographics, the need for distance education and virtual classrooms will expand dramatically. We must ask, How can the essential factors for teaching and learning available in a physical classroom be adapted and improved upon for virtual classrooms?[3]

Physical classrooms offer the equivalent of wide-bandwidth experiences. Students can interact in real time—using natural stereo sound and full-motion vision—with all the other people involved in the event. With current technologies, this is difficult to simulate. Video conferencing, virtual reality, and Internet-based synchronous communication devices (Internet Relay Chat or multi-user dimensions) are all either unwieldy, expensive, highly limited, or underdeveloped. Tiffin likens our abilities to the first automobiles: although they were revolutionary and provided the feeling of auto travel, they were impractical and barely usable. So it is with our current communication technologies. But the necessary developments, via either satellite, cable, or telecommunications, are coming.

The future, says Tiffin, will be very different for education providers. We are near the beginning of the internationalization and large-scale commercialization of education. Trade wars among education providers will drive down prices and heighten competition (Tiffin thus echoes Drucker's analysis of our developing knowledge economy). Distance education will become the norm, the least expensive way to deliver the educational product, while face-to-face teaching will be only for the well-to-do. Tiffin compares this scenario, again, to the transition from horse to automobile. At first, horses were far less expensive to buy and maintain than were autos. Eventually, the automobile became affordable, and horses became more expensive. Now we associate horses with money: consider such expressions as "the horsey set." The problem, Tiffin notes, is that the effects of the transition to distance education are not yet known. Little evidence that this transition will enhance the educational process is available.

In an ICDE SCOP working paper addressed to UNESCO, the purposes and beneficiaries of distance education, or "open distance learning," are laid out. Open distance learning can begin to overcome the educational inequities both within and between nations. Students worldwide will be liberated "from the constraints of time and place, leading to the benefits of increased access and flexibility." The technologies supporting open distance learning may also improve the quality of education through "learner-centered" approaches "with possibilities of new ways of interaction with fellow students and other learning resources." Employers worldwide will be able to offer cost-effective, high-quality professional training in their workplaces. Governments will be able to reach underserved groups and increase the cost-effectiveness of public education.[4]

In order to achieve such comprehensive goals, distance education must be carried out through more sophisticated means than text-based synchronous computer conferencing. The National Institute for Multimedia Education in Japan, for example, provides distance education through both one-way and interactive communication involving a mix of technologies: mass media, telecommunications (satellite, fax, and so on), and packaged instructional materials (video, text, graphics). In a country where nearly all students graduate from high school and most go on to college, Japanese distance education primarily benefits the adult learner who must keep up with the pace of knowledge development.[5] A U.S. model for such a system is the National Technological University headquartered in Fort Collins, Colorado. NTU offers a variety of educational programs in engineering that originate at any one of dozens of universities and are delivered by satellite primarily to engineering professionals in business and government. Although some courses may be prepackaged and asynchronous, NTU also delivers synchronous, interactive courses. A course taught at, say, the University of Michigan may be transmitted via satellite to a remote student with a two-way hookup while it is being taught face-to-face to students in Ann Arbor.

Broadband interactive distance education is still expensive. But as computer networking software advances and as hardware continues to become faster, smaller, and cheaper, open distance learning will become pervasive. As it does, it will change the shape of educational institutions and the nature of teaching and learning. I would bet that net-savvy educational administrators and state and local legislators across North America are watering at the mouth at this very moment. They are looking at declining budgets for education, on one hand, and the possibilities of building and running larger schools with the same number of teachers or fewer, on the other. Patricia Cuocco, an administrator for the California university system, says in a *New York Times* report on the virtual classroom: "By the year 2005, we expect student enrollment to hit 500,000 [in the California university system], but we won't even have physical space on our campuses for 50,000 of those students. . . . If more class sections were offered over video, computer, microwave, satellite, there would be less competition for seats. Students would get the classes they need in more timely fashion and not languish on campus for five or six years to graduate."[6] Cuocco is speaking about a university system, but many primary and secondary school districts face the same problem: how to provide an education using insufficient resources. The promise of the net is that an investment in technology will improve the quality of education.

Of the many questions this situation raises, two jump out at me: what kind of classes would these students be offered, and who would teach them? If we go back to the ideal vision posed by Diane Ravitch, we can begin to shape some answers. Ravitch said that in the near future we will have the technology to "teach anyone anything at anytime," the AAA promise of the net (which usually incorporates an important fourth *A:* anywhere). This vision is consistent with most analyses of the net: it decentralizes information, making it ubiquitous. But while we may have expanded access to educational materials, most of those materials will be in the form of prescribed packages, which over time will tend to centralize expertise in the hands of the most efficient provider (that is, the one who supplies acceptable quality at the lowest cost). The key issue is whether synchronous interactive schools will be affordable in this new era.

*Synchronous Communication, Educational Efficiency, and the Myth of Information Transfer.* If your goal is to extend education to those who may not otherwise have access to it, you have two broad options: build more schools and hire more teachers, or deliver already existing instructional enterprises to a larger audience via communication technologies. The second approach, that of open distance learning, can be carried out in one of three ways: you can provide synchronous instruction, asynchronous instruction, or a combination of both. Synchronous instruction involves interaction between teacher and students in real time, whether through two-way satellite video hookups or through text-only computer conferences. Asynchronous interaction includes any form of prescribed instruction from printed, self-paced courses to videotaped courses to interactive CD-ROMs to multimedia hypertext programs to VR immersions.

If you provide synchronous distance education, you may indeed reach more students, but you must either hire more teachers or increase teachers' workloads and class sizes. If my employer, for example, were to arrange for me to teach a course in technical communication through, say, the National Technological University, I would be increasing the number of students I teach either by adding a separate satellite course to my current workload or by adding satellite students to my current face-to-face classes. Some administrators may think that technology lets me increase dramatically the number of seats in my forty-person classroom. As the net develops, there may be no practical limit to the number of potential students I can reach. This view would be acceptable if we believed that the teacher's job is to transfer information: the teacher possesses knowledge, students need knowledge, teacher simply transfers knowledge by explaining it clearly to the student. Robert C. Hetrick gives this advice about the new communication technologies to governing boards of universities: "Colleges and universities are really in 'the information business.' Accepting this is critical to your success, and you need to take this very seriously."[7] I believe, in contrast, that schools are in the teaching business. Hetrick and I are talking about two different jobs. If all I do is stand in front of a class or camera and lecture for an hour, I might as well reach a thousand students as forty, because all I am doing is attempting to transfer information through various conduits.

The theory of communication that underlies the information-transfer model separates knowledge from communication; it treats knowledge as an entity that exists independently of the teacher and learner. If teaching is merely information transfer, then the most important quality of the teacher is clarity: if objective bodies of information exist in ready-to-communicate states separate from communicators, then successful communication is when the receiver takes possession of that information. The key for the communicator is to make the communication channel into a clean and clear conduit through which facts can travel unimpeded. This view does not explain how information comes to mean something to a student, or why students and teachers may have difficulty communicating. When uncertainty occurs in communication, according to the theory of information transfer, the problem is not one of meaning but one of either the availability of the correct information or the channel's clarity.[8]

Teaching and learning are actually far more complex than that. They involve the negotiation, construction, and reconstruction of meanings in the minds of the participants. (Likewise, the process of reading is a meaning-generating process: you construct meanings as you read this text; to the extent that you and I share worldviews, we share interpretations of this text.) There are no clearly objective facts that proceed uninterpreted from teacher to student. Moreover, both students and teachers are changed through the process of teaching and learning. It is an engagement of minds considering the same issues, developing common languages, challenging each other's assumptions, and achieving understanding.

If I am to participate in this kind of relationship as a teacher, the number of students I can teach during any one course is limited. Synchronous communication technologies enable me to reach far more students than I can teach. And yet the hope is that the synchronous communication technologies can enhance that pedagogical relationship by making education more learner centered. A number of teachers in institutions from grade schools to colleges, for example, are using online discussions through computer conferencing systems like MUDs and MOOs to help reticent students develop their voices. In a synchronous online class session, discussion need not be monopolized by a few dominant voices because the participants can, in effect, all talk at once. The danger is that the sessions can turn into screenfuls of babble, so the job of those who design such courses is to structure them to take advantage of the multi-vocality of the net. In "Dante in MOO Space: Using Networked Virtual Reality to Teach Literature," Leslie Harris describes how he and his students used Diversity University MOO as a medium to reconstruct a fictional world — Dante's Inferno — so that college students could not only read about that fictional world but participate in it. Likewise, Merry Dykes and Jennifer Waldorf have shown how school children can learn language skills, programming skills, and social-collaborative skills through structured participation in a multi-user simulated environment (MUSE).[9] On a much larger scale, a number of high schools across New York State — in conjunction with NYNEX, the federal government, and several universities—are participating in an educational service called the Living Textbook, which enables students to engage interactively in broadband synchronous projects and presentations originating at various sites on the net.[10]

All these innovations are time-consuming enterprises that engage students and teachers collaboratively. Some become cooperative while others become confrontational, and most demand the intense involvement of the participants over time. These are not the kind of educational innovations that work well within the regime of Anything, Anytime, Anywhere, Anyone. Synchronous conferencing enables participants to connect from nearly anywhere but not at anytime and rarely with just anyone—the participants must be part of an ongoing teaching and learning process. The only way that education will achieve the technotopian dream of AAAA will be through the development of sophisticated asynchronous educational services. So, as Ravitch predicts, when "little Eva" wants to learn algebra at three in the morning, she will access the net for a pre-designed instructional package.

*Asynchronous Education, Centralized Expertise, and the De-skilling of Teachers.* For the past several years I have been a member of two interdisciplinary teams funded by the federal government through the National Science Foundation whose tasks have been to research and develop interactive multimedia (IMM) instructional packages for advanced engineering courses.[11] We were not the first to begin developing IMM for education, but we were among the first to focus on certain high-level undergraduate and graduate engineering disciplines. (My role, ostensibly that of communication consultant and instructional designer, has generally been that of skeptic.) Our stated goal has been to develop courses of study in IMM so that anyone, anywhere, anytime can access either a set of CD-ROMs or a net connection and thus take the course via a multimedia workstation.

On the surface it appears that we have all the channels necessary to match the face-to-face versions of these courses and then some: we have video and sound that can be used to simulate the presence of a teacher speaking to the student. In addition, the student can open an electronic notebook, take notes, and capture information for later offline study. But the power of the interactive, graphical capabilities of the technology is to simulate (in three dimensions where appropriate) complex technical or mathematical devices and concepts so that the student can experiment with, interact with, and play with images that in a traditional classroom may be presented statically on a slide or a chalkboard. Online, when sophisticated images are generated through powerful computational engines, the student can manipulate those images in the process of tackling problem-solving tasks posed by the program. If the topic at hand relates to the ways external stresses affect the structural integrity of an object, let's say a beam on a bridge, the student can manipulate not only a simulated beam but also three-dimensional mathematical illustrations of the interplay of forces on that beam. As VR capabilities become more robust, the student may be able to explore these images graphically, thereby participating in the math in a way that pre-immersion students may never have done. All these laboratory-like experiences can be integrated into lecture-like explanations of the material. If the course is designed carefully, students can experiment with the phenomena and get feedback at their own pace. The traditional lecture pales in comparison.

The projected trajectory of this kind of IMM development is to put course after course online until entire disciplines are housed within the technology—which, as more sophisticated graphical simulations are developed, will lead to various forms of asynchronous immersive VR. The big questions remain: How many courses can be virtualized in an asynchronous stand-alone package? Which disciplines lend themselves to these transformations? Can we anticipate all the needs of students and design these systems to satisfy those needs? If we are trying to create learner-centered education, can we do so by creating net-centered packages? Certainly a number of courses could be transformed in interesting and powerful ways, but many others exist only through spontaneous interaction of students and teachers. So, you might say, those courses will never become prepackaged; much education will remain synchronous. That is the reasonable answer, but not the likely one.

An examination of one asynchronous educational package may illuminate the choices facing us. A team headed by Jack Wilson, the director of the Center for Innovation in Undergraduate Learning at Rensselaer Polytechnic Institute (RPI), transformed its introductory physics course by eliminating the traditional lectures of four hundred students. Instead, a professor meets fifty to sixty students in a computer lab running interactive multimedia workstations through which students can proceed at their own pace. "Probes hooked up to computers let students carry out scientific experiments. Multimedia software asks questions, displays and analyzes student responses, plots results and outcomes, then asks new questions." [12] But this is not distance learning; it combines computer-generated simulations with face-to-face instruction. Even so, most people who care about quality education, I believe, would say that the RPI model is the way to go. But this experiment and others like it put fewer students in touch with each teacher. That is not an efficient way to operate in eras of tight educational budgets. As Hank Bromley notes in his analysis of the social contexts of educational computing, the economic model of cost effectiveness and worker productivity, when applied to education, leads to the increased use of computers; however, when learning effectiveness is considered, smaller classes and more teacher-student interactions would be better. "But that would mean spending more on teacher salaries, just the opposite of the economies computer advocates are promising." [13]

Given the demographic and economic troubles that plague education worldwide, the trend will be away from integrated synchronous or asynchronous schooling and toward the portable, the flexible, and the preprogrammed. The technotopists bill this as an advance, which it is in efficiency but not in effectiveness. As John Tiffin predicted, competition within the educational marketplace will become fierce, winnowing out the education providers that cannot compete in quality and price. Teachers will become less central to the learning process. Michael W. Apple

argues in his analysis of the de-skilling of teaching via computer technology, "Instead of teachers having the time and the skill to do their own
curriculum planning and deliberation, they become the isolated execu-
tors of someone else's plans, procedures, and evaluation mechanisms."[14]
Further, while the number of access points to learning will proliferate, the
number of providers will shrink. Concurrently, preprogrammed educational packages will, by definition, limit the kinds of education available.
Rather than an era of pedagogical plenty, it will be an era of *connectivity
plenty* in which everyone can connect to an ever more centralized core of
expertise.

The excellent work of those like Jack Wilson at Rensselaer highlights
by contrast the impoverished and corrupt nature of the traditional American system of higher education. It is impoverished because it is based
on an information-transfer model of teaching: pack the students into the
lecture hall and let the information wash over them. It is corrupt because
it is based on a reward system that engenders faculties of career researchers who, in order to maintain their research agendas, must teach as little
as possible, thereby requiring large classes. Net-based distance education
will free up time for researchers to chase dollars, to build reputations for
the institutions, and to embellish the star system further while the peda-
gogy is centralized and homogenized and teachers become facilitators.

The ideal is to put students into small classes with skilled, dedicated
teachers, supported by communication technologies appropriate to the
area of study (and sometimes books are the most appropriate). But the
ideal has always been a luxury. The gap between the educational haves
and the have-nots will widen, not decrease, with increased net access.
Ironically, when we talk nowadays about educational have-nots, we are
talking about schools and students who have no access to the net. As the
net becomes ubiquitous, the real problem will be limited access to face-
to-face teaching and learning environments.

The issue hidden in all the visions of open distance learning is the relation of schools to local places. The virtualization of school removes it

from the fabric of the local community. Add to that the inevitable virtual-ization of work and most forms of entertainment (and religion, and poli-tics, and so on), and what will be left of local life? The dominant image in the age of the net will be the nomad. If one has no need to be anyplace, one has no place. Will local communities go the way of the horse and be-come available primarily to those with abundant disposable capital? I can imagine a new industry of professional role-playing community mem-bers hired by the super-rich to serve as their neighbors in Simulacra City. In that world I hope to play the seedy guy at the end of the simulated Dunkin' Donuts counter who blathers on and on while no one listens.

# Part Four

## The Wired

## Neighborhood

Culture rises from geography because *place* is a primary shaper of the soul. Place, as I am thinking about it, is the character of a particular landscape as altered by human occupation. Culture is what happens when individual souls find themselves gathered and interacting in this place.

—*Gregory Conniff, "Where Do You Love?"*

# 9 The Communitarian Vision

What binds individuals into a community is often apparent when that community ruptures. Picture a peaceful night a few years ago in a farming community in the Amish country of the Kishacoquillas Valley, a stretch of farmland sitting within an expanse of wooded mountains and fertile valleys that cuts across central Pennsylvania. At one moment the night was peaceful and dark. The next, it exploded. First one barn burst into flames, then another. The arsonist, Darvin Peachy, was on the move, igniting his targets across the valley. Livestock were immolated. Horses bolted from their stalls only to burn to death in the firestorm of falling timbers. In a few hours entire herds of dairy cows and the livelihoods of their owners were destroyed. By the time Peachey had finished, six barns had been ruined and two damaged, and more than 175 animals had been killed. All the burned farms were Amish farms.[1]

Why would someone attack the Amish community, whose members are so notably nonviolent? Was this conflagration motivated by an outsider's prejudice against the sect? Or had tensions within the community, invisible to outsiders, revealed themselves through this spectacle of public violence? These questions were answered in part over the next two years as Darvin Peachey was arrested, tried, and convicted of arson. This crime, it turned out, exposed a legacy of antisocial behavior that had evolved within the Amish community. Darvin Peachey was the descendant of both an arsonist and the Amish. His father, Abraham Peachey, had been convicted of arson years before and in 1965 had served a three-month sentence for burning down a barn where he worked. Although Darvin and Abraham Peachey were not practicing Amish, Darvin's grandfather was a bishop in the sect.

In a published interview given more than a year after that night, Samuel Yoder, one of the arson victims and a neighbor of Bishop Peachey, commented on the nature of the bond among the people of his community. " 'I can hardly express myself,' said Mr. Yoder, who says he is still reeling from the disaster, the outpouring of help and the arrest of Bishop Peachey's grandson. And while he would not go so far as to say that he was glad that someone had burned down his barn, he did say that the experience strengthened his faith. 'It gets you stronger,' he said. 'It makes you want to get out and help other people.' " [2] As a testament to the productivity of the community, all the barns were rebuilt within a month of the fires.

In this case the bonds of a common faith and a common way of life served as a buttress against the anomie and sickness that can grow both within and outside those bonds. But sometimes a rupture itself is, in a destructive way, a community. Picture an inner-city gang whose lifeblood is selling drugs:

> Chill said the gang cared about its members like a family would and kept them in line: "They told us we can't even sell drugs around here if we don't go to school. I know one brother who was broke all summer because he messed up in school and couldn't sell."
>
> Chill said the boy improved his school record and was back on the corner selling.

"It's no different than your mama making you go to school," Chill said.

"And anyway," Chill added, "a lot of mamas don't care if you go. The organization does. It don't want no dummies. The organization ain't all negative." [3]

Here is a community so stressed, so broken, that what counts for communal care and compassion is also the source of its inevitable dissolution: its ties do not bind but choke.

In both cases relatively homogeneous groups are constrained by their environments. What makes communities out of these collectives of individuals cannot be separated from the places where they live and the people they live with. Most people, however, do not experience either extreme of community; they do not live within collectives as circumscribed as Amish sects or as self-destructive as drug gangs. Most people live in cities and towns in which they find too few reasons to bond either with the place or with the diversity of others who also find themselves located in that place. What threatens the well-being of most of us is not the dangers of homogeneity but the inability of individuals to coalesce across differences.

My argument thus far—that the net, in connecting everyone, furthers our isolation by abstracting us from place and virtualizing human relations—will have no effect on the pace of technological development. Recall Porush's Law: "Participating in the newest communications technologies becomes compulsory if you want to remain part of the culture." Now I shall posit Steve's First Corollary to Porush's Law: "If you want to enhance the culture, steer your participation in the net toward ways that better integrate you and others into your local geophysical communities." (Steve's Second Corollary: "Be wary of the seductions tendered by the immersive virtualists.") Given the inevitability of the net, the most fruitful path is to participate in it in ways that benefit our localities. By taking this position, I am accepting the arguments posed by a number of social and technological activists I call the wired communitarians: those whose agendas are carried out primarily through a loosely connected group of community-based computer networks. These nets come in a variety of sizes and configurations, from the expansive urban National Capital Free-Net in Ottawa, Ontario, to the mid-sized Blacksburg Electronic Village in rural Virginia to the tiny school-district-based North Country Network under development where I live.[4]

This wired-community activism has not suddenly materialized out of the ether. It is not merely the product of computer network technologies. Its origins can be traced back through earlier technologies and periods of techno-utopianism. Even before personal computers, many believed that some form of communication technology could erase the isolation and alienation of the individual.

Yet when I picture isolated and alienated individuals, I can't help but envision images conjured up by Chill's story about a community in which crucial bonds like those among mothers, children, and schools are weakened or undermined by destructive forces. I feel compelled to hold up such images as a way to measure the aims and achievements of the wired communitarians. I'm thinking about the children I see on my way to work waiting for the school bus in front of rotting, listing houses. I'm thinking of the elderly in a poor Chicago neighborhood who are afraid to open a window during the worst heat wave in decades. I'm thinking of a mother in the Old North End of Burlington, Vermont, a mother who has never touched a computer, who has never used a fax machine, who may not have a telephone, whose use of broadband communication technology begins and ends with television, who cannot type, and who has few, if any, technical skills. How can she and her children become better connected to the opportunities in her community? How can she become touched by the wired communitarians' movement?

The range of activities commonly referred to as community networking is quite diverse. Consequently, as observers and activists are beginning to write the history of these activities, they are shaping a whole that may be larger than the sum of its parts. And that whole is beginning to look like a populist communitarian movement. As such, the evolution of community computer networks is often used as an example of the law of unintended consequences. Following the development of ARPANET, the proto-Internet that connected federal and university laboratories dedicated primarily to conducting military research, a handful of small, unrelated bulletin board systems and local area networks began to take shape across the country. As the chroniclers tell it, technologies like the ones developed through large-scale, federally funded, hierarchical programs were somehow being appropriated by energetic visionaries who saw the need for people to be better connected to one another and found a new means to connect them—people like Dave Hughes and his political- and community-oriented Old Colorado City bulletin board system; Frank Odasz and Big Sky Telegraph, which attempted to network tiny, isolated rural schools and communities in Montana; Tom Grundner and his family health care bulletin board, which evolved into the large Cleveland Free-Net; and Ken Phillips and the Santa Monica PEN, a community net designed to promote participatory democracy on the local level.[5]

These enterprises and the many other community nets that have grown up around them may differ in a number of ways, but they all originated locally to fulfill the needs of the community—or at least that is the vision. Says Tom Grundner:

> America's progress toward an equitable Information Age will *not* be measured by the number of people we can make *dependent* upon the Internet. Rather, it is the reverse. It will be measured by the number of *local* systems we can build, using *local* resources, to meet *local* needs.
>
> Our progress will not be measured by the number of college educated people we can bring online—but by the number of blue collar workers and farmers and their families we can bring online.
>
> It will not be measured by the number of people who can access the card catalog at the University of Paris, but by the number of people who can find out what's going on at their kids' school, or get information about the latest flu bug which is going around their community.[6]

Other leaders in the field concur:

Mario Morino and the Morino Institute say that community nets must be forums for local public deliberation in order to enhance local self-determination. They must serve to organize local information and human communication. They must be used for the good of the less fortunate in a community: the low- to middle-income families, the disabled, the immobile. They must provide affordable—possibly free—access for all. Most important, they must do "what commercial providers find difficult to do well: represent local culture, local relevance, local pride, and a strong sense of community ownership."[7]

In a report to the Rockefeller Foundation on the revitalization of local communities in the face of community disintegration nationwide, the Millennium Communications Group argues that organizations like community nets can function as "small scale community institutions that become vehicles for action, the incubators for solutions, and the metaphors for enablement."[8]

According to Doug Schuler, chair of the board of directors for the Computer Professionals for Social Responsibility, the politics of community networking should encompass five goals: community cohesion, informed citizenry, access to education and training, strong democracy, and an effective process through which these goals can be achieved. The kind of group that develops a community net, says Schuler, is an "*ad hoc* alliance of librarians, educators, network and bulletin board systems users, community activists, social service providers, government agencies, and concerned computer professionals."[9]

Miles Fidelman, the executive director of the Center for Civic Networking, campaigns for community nets that play a role in the sustainable development of local communities. Fidelman and others who believe that "economic development, environmental protection, and quality-of-life are all inter-related and must be planned for on a long-range basis" are developing SDINs, Sustainable Development Information

Networks, in a variety of locales. To focus on the needs of an entire community, a diverse set of community members, including representatives of such opposing groups as local businesses and environmentalists, must be involved in the work. Overall, SDINs will serve as a repository for local statistical, demographic, and geographic information; as a tool (eventually including simulation capabilities) for organizing and manipulating local information; as a tool for accessing relevant regional and national statistics, demographics, and geographic information and for contributing locally generated information to regional and national planning activities; as a way to explore what other communities are doing and to share experiences with peers in other communities; and as a vehicle for communication and collaboration concerning the support of local activities and of local participation in regional and national activities.[10]

Richard Civille, also of the Center for Civic Networking, characterizes the overall movement as the development of community information infrastructures and argues that only through local efforts can the so-called National Information Infrastructure develop.[11]

In her analysis of the words and actions of such community net activists as these, Anne Beamish identifies five assumptions about the nature of network technologies that inform their work: (1) the new communication technologies will be increasingly important to economic and social development; (2) these technologies "reinforce more democratic and decentralized modes of communication"; (3) they have the potential to "reinforce face-to-face patterns of communication"; (4) as an infrastructure they are more like public utilities than private commodities; and (5) they will remain desirable "despite rapidly changing technologies and policies."[12] These assumptions, hopeful as they are, underlie a social agenda that requires the technologies to be employed to serve, not to transcend, localities.

As this movement proceeds, the obvious questions arise: Will the movement succeed? Can it sustain itself? Will it deliver what its advocates claim it must deliver? The answers are not yet clear, nor will they be for some time. But the movement has achieved tremendous tangible gains in a relatively short time. What remains to be seen is whether the early successes will lead to the communitarian goals of the movement's visionaries.

All we have to do is look at the Free-Net movement and its evolution into the National Public Telecomputing Network (NPTN) to see how quickly community nets are spreading. According to NPTN's own history, in 1984, while working in the Department of Family Medicine at Cleveland's Case Western Reserve University, Tom Grundner constructed a computer bulletin board system to test how well telecomputing technologies could deliver health care information to the local community. This system, named St. Silicon's Hospital and Information Dispensary, ran on one phone line. Anyone with a computer and modem could connect to an interactive service and write a question about family health care. The question would be answered within twenty-four hours by a board-certified family physician. The experiment worked and eventually attracted major university and corporate support, which was used to expand the service.

In the summer of 1986 the newly named Cleveland Free-Net opened, providing not just medical information but locally generated information relating to the law, education, the arts, the sciences, and the government. Over the next couple of years the system attracted more than seven thousand subscribers and handled between five hundred and six hundred connections per day. By 1989 the Free-Net had migrated to a more powerful computing system and significantly increased its capacity.[13] Within five years of this upgrade the system had more than a hundred thousand subscribers, handled more than fourteen thousand connections per day, and hosted more than 120 special-interest groups contributing a range of local information.[14]

Along with expanding the Free-Net's capacity, Grundner and his associates formalized a process by which they could spread their technology and expertise. In 1989 they established the National Public Telecomputing Network (NPTN) to help other communities develop similar networks, to link those local nets, and to provide services, features, and programming (known as cybercasts) to all member networks. Over the next five years dozens of new Free-Nets were initiated across the United States and Canada, and around the world. By the spring of 1995 there were more than 50 NPTN-affiliated community networks in operation and 120 in development, located in 41 states and 10 countries (Australia, Canada, Finland, Germany, Ireland, Italy, New Zealand, the Philippines, Sweden, and the United States), with most in U.S. and Canadian metropolitan areas.[15]

The National Capital FreeNet (NCF) of Canada's Ottawa-Hull region is probably the world's second-largest community net. Its development illustrates the pace at which these enterprises can proceed. In November 1991, several people from Ottawa's Carleton University met to discuss the possibility of developing a Free-Net based at Carleton's computing services department. The first public meeting to begin planning the effort was held in March 1992, attracting a hundred people from fifty organizations. The NCF officially opened on February 1, 1993, and by May 1995 had more than forty-two thousand registered users and received information from more than 250 organizations.[16]

When you connect to the NCF, you see a menu like this one:

<<National Capital FreeNet—Main Menu>>
 1. About the National Capital FreeNet . . .
 2. Administration . . .
 3. Post Office . . .
 4. Public Discussion . . .
 5. Social Services, Health, & Environment Centre . . .
 6. Community Associations . . .
 7. The Government Centre . . .
 8. Science, Engineering and Technology Centre . . .
 9. Schools, Colleges and Universities . . .
 10. The Newsstand . . .

11. Libraries . . .
12. Special Interest Groups . . .
13. Communications Centre . . .
14. Professional Associations . . .
15. Help Desk . . .
16. Menu principal français . . .
17. Ontario Provincial Election Project

---

h = Help    x = Exit FreeNet    p = previous    u = up    m = main

These menus lead to hundreds of choices, enabling users to retrieve archived information, to participate in asynchronous discussion lists and synchronous chat groups, and to maintain e-mail accounts. If you choose item 1, you find a description of the system's operating philosophy:

> The National Capital FreeNet is a computer based information service designed to meet the present information needs of the people and public agencies in the region, and to prepare the community for full and broadly based participation in rapidly changing communication environments. . . . .
>
> By dialing into a single number, members of the community have a free, twenty-four-hour connection to information and interaction with any of the participating organizations. . . . The FreeNet is an electronic community centre, public square, and information fair.
>
> Rather than each community agency trying to develop and maintain its own expensive, undersubscribed, single-service telecommunication service, each is part of a single, cost effective, sophisticated, multi-line computer facility. The FreeNet is, then, a shared platform with each organization having a large, dedicated space on the system and autonomously determining how it is used.

If you choose item 6, Community Associations, you will find dozens of organizations and agencies with dedicated space on the net, such as the following:

Citizens for Safe Cycling
Ottawa South Community Association

Scouts Canada, National Capital Region
Humanist Association of Ottawa
Riverside Park Community and Recreation Association
Orleans Little League
Canadian Kennel Club
Baha'i Faith Forum

If you choose item 12, you will find many local and regional special-interest groups (SIGS), such as these:

Christian Youth SIG
Disability SIG
Gay, Lesbian and Bisexual SIG
Parenting SIG
Seniors SIG
Contract Bridge SIG
Role Playing SIG
Lateral Thinking SIG
Librarians and Information Professionals SIG
Rehabilitation SIG
Translation/Interpretation SIG

I usually connect for one of two reasons. Sometimes I am collecting information about how the NCF operates. I may want to see how many people are choosing different menus. I may want to read the discussion list of the board of directors, or I may want to find out when and where the next public meeting will be held to discuss the net's progress. Other times I choose item 10, The Newsstand, to see what is happening in the Ottawa-Hull region over the weekend: concerts, plays, community events, and the like. I should point out that even though I visit Ottawa frequently with friends and family, I am not a typical subscriber because I telnet to NCF from my host system. Most users connect either by using one of the dozens of public access terminals located in nearly twenty locations around the Ottawa metropolitan area or by dialing one of the hundreds of phone lines provided by the NCF—lines that are continually busy. And therein lies one of the indicators of a popular system: the demand for its use always seems to outstrip its capacity.

Although it makes sense that the largest Free-Nets are located in metropolitan areas, there is growing support to develop such enterprises in rural regions, as well. In 1994 the National Telecommunications and Information Infrastructure Assistance Program of the U.S. Department of Commerce awarded the NPTN nine hundred thousand dollars to support the development of rural community nets. Rural regions need the support. In the spring of 1995, of all the NPTN affiliates in operation or in development, only ten were identified as rural nets.[17] But not all community nets, rural or otherwise, are NPTN affiliates. The net closest to home for me, the North Country Network, is a non-NPTN net under development by a coalition of regional school districts, public libraries, local universities, and a local phone company.

The places served by "my" two nets couldn't be more different. The home of the National Capital FreeNet is a cosmopolitan place, vibrant and diverse. Besides serving as the heart of Canada's federal government, Ottawa-Hull is one of the more dynamic cultural and economic centers in Canada. (I'm told that residents of Montreal, Toronto, or Vancouver might scoff at any coupling of the words *cosmopolitan* and *Ottawa;* it's all a matter of perspective.) Although the nation in recent years has suffered from persistently high unemployment and a weak dollar, Ottawa-Hull has seen the rise of a high-tech design and manufacturing sector. Among other economic strengths, it has an expanding tourism industry: the region has become a significant international vacation destination with its concentrations of museums, historic and governmental centers, and its city-wide festivals like Winterlude and the Tulip Festival. The region is also highly wired. In addition to the connections available through the area's major universities and through the FreeNet, residents have a choice of a number of commercial Internet access providers.

Less than sixty miles to the south of Ottawa lies St. Lawrence County, New York, the primary setting for the North Country Network. Encompassing twenty-eight hundred square miles, the county is the largest in the state, larger in area than either Rhode Island or Delaware, but it is also New York's least populated county, with only 114,000 residents. Further, it is consistently one of the three poorest counties (out of sixty-two) in the state; it has one of New York's highest unemployment rates, lowest per capita incomes, and highest percentages of people with incomes below the federal poverty level. The economy has been dominated by an agriculture industry that has lost nearly a third of its farms over the past twenty years.[18] Finally, the region is nearly bereft of major Internet access providers—except for those available through the universities. All the big American net companies, like CompuServe or America Online, can be accessed only by long-distance telephone calls. In 1995 one company with six phone lines was selling Internet access connections—primarily to a few businesses—and another such company was forming. But adequate net access will not be a problem; a vibrant community net will.

Urban areas have the critical ingredients to support community nets, but much less evidence is available to show that remote rural areas can also support them. A particularly successful rural community net—albeit one serving a much more densely populated region than New York's north country—is the Blacksburg Electronic Village (BEV). In 1995 more than a third of the thirty-six thousand residents of Blacksburg, Virginia, participated in the net. The BEV, which connects people to schools, the government, health care providers, businesses, and each other, was created through a joint effort by Virginia Polytechnic Institute, Bell Atlantic, the city, and a number of local organizations and businesses.[19]

Early in its development, BEV elicited great hopes among Blacksburg residents. A 1993 study of residents' attitudes toward BEV conducted by Ann Bishop and Scott Patterson reports that although most users were relatively new to the system, they had enthusiastic visions of its future impact on Blacksburg: "The greatest excitement seemed to stem from the possibilities of developing local resources on BEV and attaining local ubiquity of the system; the desire to use BEV to enhance the town's sense of being, and functioning as, a community was very strong." Users saw the system enabling them to make new connections among the townspeople; they saw it empowering people in the community, especially those who were "physically and psychologically limited in their abilities to interact with others." In addition, users envisioned BEV helping them to tap into "local information for civic or cultural purposes": to access transcripts of town meetings, community bulletin boards, university seminar schedules, and the like.[20]

Complementary results were found in a January 1994 BEV-related survey reported by Andrea Kavanaugh and Scott Patterson. When asked about potential net services, most respondents expressed "some or much interest" in areas that affected their community lives:

83 percent had some or much interest in local events,
79 percent in local news,
78 percent in online emergency (health) information and help,
68 percent in "ask a nurse or doctor" information,
67 percent in prescription drug information,
60 percent in home security services.[21]

Both surveys indicate that the BEV users expect the net to do what the civic activists and community net proponents hope it will do: reinvigorate the geophysical community.

How might this occur? According to BEV's "View of Blacksburg of the Future," the wired village would include services like a "local research and educational network" that would, among other things, enable students to collaborate with their peers at other schools in other towns near and far, to conduct research via net archives around the world, and to take "electronic field trips" via switched video technologies. At the same time this educational network would allow for completely local applications. Instruction could be delivered electronically between home and school.

Library services could be accessed from home and from remote class-rooms at any time. Through an "electronic buddy system" students could keep in touch more easily with mentors from the community. Parents could more easily monitor their children's progress by accessing course materials and engaging in regular electronic communication with teach-ers, mentors, and administrators.[22]

In addition, the organizers of BEV envision electronic town meetings and better delivery of certain local health care services, increased local commerce, and secure financial services via the network. And they also emphasize that the net will improve the quality of life:

> In Blacksburg, electronic mail is becoming as ubiquitous as making a phone call or receiving mail from the U.S. Postal Service. Situations that require getting in touch with an entire group quickly, such as the postponement of a play rehearsal or a team practice, can use the mailing list feature of electronic mail to distribute the message to everyone.
>
> Electronic bulletin boards (Usenet) and electronic conferences (mailing lists) are being conducted on topics as diverse as senior activities, where to repair your car, and youth soccer leagues.[23]

But to believe that all these kinds of applications are feasible is to assume that a community is populated by educated, connected, and computer-literate people. The distinct possibility exists that community nets may only continue to network those already networked by other means, those already engaged in their communities, those already participating in local civic issues, in their children's education, and in the local economy. Will the North Country Network help to improve the economic, social, and educational conditions of poor, rural St. Lawrence County so that those not already networked will gain the means by which to become better connected? Will civic nets have any effect on the low-skilled, unwired family in Burlington's Old North End? Can the big urban nets, like the NCF, reach across the diversity of their cities?

A 1995 survey of the users of the National Capital FreeNet conducted by Andrew S. Patrick, Alex Black, and Thomas E. Whalen indicates that net users do not simply come from one socioeconomic segment of the local population. But neither do they match a cross-section of the region. The authors compare the demographics of the survey respondents to those of the region in five of the six following categories:

*Gender:* Approximately 80 percent of survey respondents are male, whereas 49 percent of the region's population is male.

*Age:* In comparison to the region, the NCF under-represents the young (under age fifteen) and the old (over sixty-five) and over-represents those between fifteen and nineteen and between twenty and twenty-four.

*Education:* NCF users "differ most from the local region in the category of university educated . . . with more than 50% of the NCF users being university educated. . . . There were also fewer NCF users with [only] a grade school education than might be expected from the regional data."

*Occupational Status:* The survey indicated that approximately 40 percent of users were salaried employees and nearly 30 percent were students, but the only comparable regional data dealt with unemployment. The respondents showed slightly less than a 4 percent unemployment rate, while the region three years earlier had shown a rate of more than 7 percent.

*Income:* "The comparison with the regional income levels shows that NCF users are both better off than the local community (incomes greater than $50,000) and less well off (incomes less than $9,000). There is a corresponding under-representation of people in the middle income ranges ($10,000–$40,000)."

*Computer Experience:* The average amount of experience was approximately eleven years, with most NCF users having between six and fifteen years of experience.[24]

Throughout their report on these results the authors argue that the data reveal a demographic reach greater than expected and certainly greater than that of several other large, urban community nets. Is this evidence that community nets might be able to achieve the communitarian goals articulated by net activists? It is probably too soon to tell. What is clear, however, is that increasing the range and diversity of net users is only one of a number of challenges that face the community network movement now and in the near future.

## 10 Challenges to Community Networks

In her analysis of the community network movement, Anne Beamish identifies several short- and long-term goals of community nets. In the short term the nets must achieve growth and sustainability. That is, they must develop staffing and funding mechanisms that enable them to continue operating, and marketing mechanisms that attract members and information providers. In the long run, says Beamish, the nets must somehow facilitate community-wide access, generate public debate and democratic participation, and encourage community development.[1] These goals fit into two broad categories. The first, encompassing growth, sustainability, and access, has to do with the infrastructure of community nets; the second, concerning democratic participation and community development, involves their activist agendas.

*Infrastructure: Growth, Sustainability, and Access.* Since 1989 the number of nets, the number of registered users, the range of services, and the amount of information made available have exploded. Take the state of Ohio, where Free-Nets and the NPTN began. By the spring of 1995, most of Ohio's population had the opportunity to connect to a local or regional Free-Net. Seven metropolitan and two rural areas had nets in operation, and five more areas had nets under development:

NPTN Affiliates (Nets in Operation)
  Metropolitan Networks
    Southeast Ohio Regional Free-Net—Athens
    Tristate Online—Cincinnati
    Cleveland Free-Net—Cleveland
    Greater Columbus Free-Net—Columbus
    Dayton Free-Net—Dayton
    Lorain County Free-Net—Elyria
    Youngstown Free-Net—Youngstown
  Educational Networks
    Learning Village Cleveland—Cleveland
  Rural Networks
    Richland Free-Net—Mansfield
    Medina County Free-Net—Medina
NPTN Organizing Committees (Nets under Development)
  Metropolitan Networks
    Akron Regional Free-Net—Akron
    Wood County Free-Net—Bowling Green
    Stark County Free-Net—Canton
    Lima Free-Net—Lima
  Rural Networks
    Tuscarawas Valley Free-Net—New Philadelphia[2]

The growth of community nets in Ohio may presage a wave of such nets over the rest of the planet. But establishing and maintaining a community net requires a lot of volunteer effort by skilled community members. Once a community net is established, finding subscribers and information suppliers is not the most difficult problem; the far harder task is paying for the net's continuing operation. Steve Cisler, a community net activist, notes, "Many of the present community networks are labors of love; they draw on the volunteer spirit of both technical and non-technical citizens in a town or region . . . but the day-to-day activities and the financial burden of growing a system to meet the demands of an ever-expanding base of new users can try the unity of even the most energetic and cooperative organizing groups."[3]

The NPTN has advocated the development of a Corporation for Public Cybercasting (CPC), modeled on the Corporation for Public Broadcasting, which would operate as a taxpayer-supported source of funds for NPTN community nets. But U.S. taxpayer support for the CPB and public television has waned, and the prospect of the federal government's backing another such enterprise is unlikely. Further, as Cisler points out, the telephone companies have tremendous political power, and some of them consider community nets their competitors. The most notable of these has been NYNEX, the Baby Bell that serves New York and New England. In response to a 1994 article in the *Chronicle of Higher Education* entitled "Some Commercial Internet Services Object to Tax-Supported Projects," Tom Grundner wrote:

> I was astonished to read the comments of Jeffrey W. Ward from NYNEX Corporation, who said: "We do not want government funds going to support or promote or bias commercial activities. What that means . . . is if a university hooks up a FreeNet, and the university hooks up the local shoe store, the local county government, perhaps the local hospital, and charges them some money to help recoup the costs, we think that's bad public policy and a waste of taxpayer funds." . . . First, it might be gently pointed out that it was taxpayer funds that paid for the development of the Internet in the first place. It is taxpayer funds that pay for much of the backbone right now. It will be taxpayer funds that will be paying for many of the coming improvements. . . . Because of the Free-Nets, for the first time the average taxpayer is receiving a trickle of benefit from these expenditures and suddenly that is "bad public policy and a waste of taxpayer funds." Go figure.[4]

Other telephone companies, however, have shown solid support for community nets. Bell Atlantic supports the Blacksburg Electronic Village, and US West has supported Big Sky Telegraph. Some of the remaining small independent telephone companies support community nets, too. Although the North Country Network lies within the purview of NYNEX, for example, the tiny, flexible Nicholville Telephone Company also has twenty-one hundred regional customers and will provide net access for all the schools and businesses it serves.[5]

Because community nets are nonprofit public access instruments offering free (or very cheap) connectivity, they need to treat carefully any revenue generation that might compete against the commercial net access companies. The National Capital FreeNet, for example, does not sell advertising space. Instead, it recognizes sponsors—a subtle but significant distinction. By noting in its opening screens what organization is sponsoring the connection that day, the NCF provides visibility for a patron while retaining control over the patron's message. As an alternative, the NCF may eventually include "ad pointers" at appropriate spots. These notes or links would point to commercially available electronic advertising services—for a fee paid to the NCF.[6]

Overall, the NCF has developed a sophisticated strategy in the face of daunting needs. In 1994, it raised more than 250,000 Canadian dollars. In 1995, the director of development estimated that the net needed to raise well over twice that amount to "ensure that the system can be accessed by the thousands who are knocking at our door, eager to be a part of the information society." The NCF targeted several groups for financial backing: NCF members; NCF directors; NCF information providers; nonprofit associations; SIGs; municipal, regional, provincial, and federal governments and government agencies with goals in common with NCF, including community development, community service, greater communication and understanding, and economic development; and, finally, corporations and businesses that are not information providers but can benefit from the success of NCF.[7] In short, a broad-based community net must continually encourage broad-based financial support. Community nets thus must compete against all the other local causes engaged in the never-ending campaigns for funding.

One of the driving forces behind the growth in community-net budgets is the apparent need to keep upgrading the quality of connectivity.
As Steve Cisler puts it, "How do you keep 'em down on VT100 after
they've seen Mosaic? The challenge of serving all users, whether they have
a Commodore 64 or a PowerMac—or no computer at all—is complex
and expensive. . . . Many systems are settling on a text-based system for
VT100 terminal access as a low common denominator. . . . some of the
high-end users may be rapidly bored and move on to systems (free or for
profit) that offer a richer set of options and interfaces."[8] The second half
of Cisler's question concerning VT100 is already dated (how about "after
they've seen three-dimensional virtual reality markup language [VRML]"?
or "after they've seen the CAVE"?). Those who run community nets have
to answer the question, Can—or should, or must—community nets attempt to evolve with the technology?

One potential answer to this question has been posed by a group of
community net developers calling themselves the Telecommons Development Group (TDG), based in Guelph, Ontario. TDG, claiming to be the
first workers' cooperative for the development of community networks,
has designed a system that enables communities to, in effect, hire out the
development and maintenance of the technology while maintaining control over its design and use. Each subscriber to this system, called Free-
Space, has a vote in the way the system is operated. The technology will
be used to run a number of community nets even though each community will build and design its own net locally. Because of the efficiency of
this system, TDG maintains, technology deliverers, or "service providers,"
like TDG can offer each community net a range of technological sophis-
tication. The co-op system also allows the service providers to compete
in commercial markets while focusing on the development and maintenance of community net functions:

> Because the central computer facility services a number of interdepen-
> dent yet independent electronic community networks, its user base is much

greater than that of a single electronic community network. Think of it as a regional newspaper with a much larger circulation than the average local community paper—yet [it] is still customized for each specific community. The large number of users accessing community FreeSpace services allows for the development of at least 4 ways in which the service provider is able to generate revenue and sustain free public access to their FreeSpace. They are, briefly, the involvement of business through customer service bulletin boards and forums, advertising and virtual transaction; value-added services such as newsfeeds, online arcades, etc.; commercial-oriented training workshops; and fees for extended user access.[9]

But the central purpose will always be community networking; therefore, each community with a FreeSpace-based net will have at least one public access computer lab to serve those without access to a computer and to function as a meeting place for local training and net-based educational programs.

In Wellington County, Ontario, the first FreeSpace is under development. According to the "Wellington FreeSpace Mandate," the FreeSpace will "plan, develop and implement programs, policies, and procedures that will endeavor to deliver barrier-free access to communication technologies in an effort to enrich the economic, cultural, and educational lives of Rural and Urban residents of Wellington County."[10] In doing so, the Wellington FreeSpace will attempt to overcome disparities in access to areas within the urban centers and between the urban centers and the rural areas. The overall goal is to provide free access to more than two hundred thousand residents of Wellington County.[11]

Although the number of individuals with access to the Internet has been rising dramatically, only a small percentage (well below 10 percent) of Americans are regularly online. As Stephen Miller of the Computer Professionals for Social Responsibility points out, "as a result of price increases caused by deregulation, a growing number of Americans—up to 20% of some low-income communities—don't even have home telephones." The state of affairs is so bleak, notes Miller, that even if we assume that connectivity will become as easy to obtain as a television or radio, universal access will not be the answer to our needs. Because no matter how cheap and ubiquitous the technology becomes, no matter how easy it is to use, in a country with a 40 percent high school dropout rate and a 20 percent adult illiteracy rate, universal communitarian participation in the net will not happen. At best, Miller says, many people will not be able to "do more than the most basic types of (probably consumption-oriented) activities."[12]

As enterprises like FreeSpace illustrate, issues of growth and sustainability cannot really be separated from those of access. Further, access cannot be separated from the social agendas of community nets. Although the difficulties of providing access to community nets is and will remain significant, pervasive access to larger-scale nets will evolve. The problem is that the nets to which everyone has access may bypass the communitarian agendas that give community nets their identities. So for those who promote community networking, the question is not "Can we achieve universal access?" but "Can community nets survive in an environment in which most people are connected to worldwide nets?" With this question, the issue of access shifts from being one of infrastructure to one of public participation and social transformation.

*Activists' Agenda: Democratic Participation and Community Development.* Some community nets are experimenting with new forms of online political deliberation in order to expand democratic participation. In 1993 the Victoria FreeNet (located in British Columbia) and the National Capital FreeNet cooperated to organize electronic political forums during that year's Canadian federal election. The nets enabled candidates to post campaign information and created two discussion groups: a newsgroup to discuss the national election spanned both FreeNets, while the NCF hosted a newsgroup devoted to local elections. By October 1993 the two groups had garnered more than 680 postings (e-mail messages), and near the end of the campaign about two hundred users a day were viewing the newsgroups. But only a small number of candidates registered as users and engaged in the discussions. "Thus, we were successful in providing useful information to voters and in providing a forum for exchanging opinions," said Richard Taylor in his report to the NCF on the project, "but we failed to achieve the two-way communication between candidates and voters that we had hoped for." [13]

In the fall of 1994 the Minnesota Electronic Democracy Project held what it called the first online debates, during the campaigns for the U.S. Senate and for the governor of Minnesota. The debates, run by the Twin Cities Free-Net, the League of Women Voters, and the Minnesota Regional Network, were part of a larger project that created online space for a variety of information concerning the 1994 election, including Minnesota candidates' position papers and an online discussion forum, a "proto-type for the 'electronic town-hall.'" This enterprise was, according to project documents, a manifestation of the "vision for invigorating the democratic process through the use of computer mediated communications." The debates were held through asynchronous discussion groups: each candidate was sent the same three questions and was required to write responses during a prescribed time. As the responses were posted, the candidates could rebut their opponents. By one week after the election, the campaign-related information had been accessed more than forty thousand times.[14]

The designers of both these projects are attempting to develop virtual agoras, electronic spaces in which candidates and citizens can speak (textually) about the issues and the election. But such exercises may be about not just elections but public engagement—about individuals facing each other and hashing out their differences. And when a community net encourages public discussion, it must have clearly drawn policies regarding acceptable and unacceptable uses of the system, the extent of users' privacy, and the power the system administrators have to censor users' discourse. Such an "acceptable use policy," as described by NPTN's Peter F. Harter, "is a contract between the user and the other users of the network, overseen by the system administrator. This contract connects each user together; this contract represents the ties that bind individuals and institutions together to form their electronic community. By agreeing to a common bond, users of a system are letting this contract be their collective conscience." [15]

In the NCF, a member's online rights and responsibilities begin with the agreement each prospective member must sign in order to become a registered user. In its membership agreement, the NCF makes two statements that describe its power to censor and the extent of its members' privacy. Statement 1 indicates that the NCF Board can censor as it sees fit by removing an abusive or fraudulent member from the net:

> 1. That the use of the System is a privilege which may be revoked by the Board of Directors of the System at any time for abusive conduct or fraudulent use. Such conduct would include, but not be limited to, the placing of unlawful information on the system, the use of obscene, abusive or otherwise objectionable language in either public or private messages, or violation of this Agreement. The Board of the National Capital FreeNet will be the sole arbiter of what constitutes obscene, abusive, or objectionable language. [16]

The NCF Board is saddled with a heavy responsibility: depending on how you look at it, the NCF is either enacting community standards for civil discourse or exercising the power of a public institution and establishing standards for civil discourse. Either way, its reach is broad because it claims all NCF discourse as falling within its authority. Accordingly, statement 2 indicates that the NCF Board can have access to any information made available to others and stored on the net. This apparently includes private mail as well as publicly accessible files:

> 2. That the National Capital FreeNet reserves the right to review any material stored in files or programs to which other Members have access and will edit or remove any material which the Board, in its sole discretion, believes may be unlawful, obscene, abusive, or otherwise objectionable.

But the separate "NCF Policy on Privacy" statement says that the "Free-Net will take every measure possible to ensure that private information is not divulged, but it does reserve the right to examine private areas" under certain circumstances. These include "investigation of violations of the User Agreement, including offensive material or misuse of accounts," and "unauthorized scripts or programs found on the system."[17]

In an incident in the spring of 1995 the NCF did indeed protect the privacy of someone who allegedly violated the membership agreement. As described by Dave Sutherland, the chair of the NCF Board, agents for a company requested that the NCF give them the name and address of a registered user because that person, said Sutherland, "posted a method for defrauding the company to a Usenet newsgroup." The NCF did not give out the name and address but instead urged the member to contact the company agent—which the member subsequently did. In a message posted to the NCF Board discussion group, Sutherland raised the possibility that in similar situations the NCF might have legal recourse to render the requested personal information because the actions of the member would have voided the membership agreement: "It would seem that we should supply members' addresses under the situation where there is some evidence that our system has been used to break the law. . . . In some of these cases, we might judge that we should cooperate with the authorities; in others we might insist on a court order." The net's policies thus will continually be assessed in terms of legal statutes, to ensure that the net adheres to community standards for civil discourse.[18]

Community standards of censorship are also filtered through the community net. The NCF does not censor users' files or e-mail messages based on their content, according to the "National Capital FreeNet Censorship Policy." It may erase such material, but only through the systematic removal of dated files meant to be stored only temporarily on the system. The NCF does reserve the right to "cancel or remove" Usenet newsgroup postings "that seem to contravene Usenet protocol or common practice." The net also reserves the right to choose not to subscribe to certain globally available Usenet groups (such as portions of alt.sex) for, among other reasons, "controversial content."[19] As the community net grows, its discourse becomes community discourse. And thus a reciprocal relationship exists between community standards and net practices.

These policies pertain to all discourse on NCF, but the Ontario Provincial Election Project, hosted by the NCF in June 1995, stipulates further "guidelines for debate": "The discussion groups are 'un-moderated' which means that your contribution will be posted automatically to the discussion group shortly after you exit from the discussion. There is no censorship or filtering of your direct submissions to the discussion. Everything will be seen by other voters, as well as by the candidates, and quite possibly by news reporters. So please watch the language you use and keep the 'flaming' to a minimum. While certain people may have raised your ire over the last five years, let's avoid name-calling and personal insults. There is no substitute for real debate on substantive policy issues." [20] These guidelines make several comments about the nature of communities mediated by communication technologies. First, people must be reminded that they are engaged in a public forum in which other people at other computer screens are also participating. Second, anger and vitriol are common in a debate. If people accept that everyone else is angry, too, then there is no need to show anger in unproductive and offensive ways. Finally, for political discourse to flourish, participants must take responsibility for their words; therefore, the discussion group is unmoderated. Thus, the guidelines are an invitation to jump into the fray, but with self-discipline.

The Ontario Provincial Election Project was similar to the earlier projects in Canada and Minnesota. It attempted to bring online various kinds of information concerning local elections in the Eastern Ontario areas that NCF serves. When you chose item 17 on the NCF main menu, you would move to the following menu:

<<< Ontario Provincial Election Project >>>
(go ont-elect)
The election date is 8 June 1995
  1. About the Ontario Provincial Election Project . . .

2. Debate >>>
3. Information about the election process . . .
4. Information en français
5. Local Ridings and Candidates . . .
6. Ontario Political Parties . . .
7. Calendar of Election Events
Ontario-wide open discussion via the Internet:
8. USENET Newsgroup ont.general >>>
9. Make a donation to keep FreeNet free

---

h = Help   x = Exit FreeNet   p = previous   u = up   m = main

The first seven items focused on the local elections. Eastern Ontario comprises fourteen political districts, or "ridings," each of which was electing a representative. If you chose item 2, Debate, you would move into an asynchronous discussion area for all the districts. But if you chose item 5, you would see a listing of all fourteen districts:

<<< Eastern Ontario Ridings >>>
1. Carleton . . .
2. Carleton East . . .
3. Cornwall . . .
4. Kingston and the Islands . . .
5. Lanark-Renfrew . . .
6. Leeds-Grenville . . .
7. Nepean . . .
8. Ottawa Centre . . .
9. Ottawa East . . .
10. Ottawa Rideau . . .
11. Ottawa South . . .
12. Ottawa West . . .
13. Prescott and Russell . . .
14. S-D-G & East Grenville . . .

---

h = Help   x = Exit FreeNet   p = previous   u = up   m = main

Participating in these fourteen elections were seventy-five candidates, of which only fourteen were registered users of the NCF; however, forty-three had arranged to have their campaign information posted in the appropriate spaces (to which you could move by choosing an option in the menu of districts). Three times as many candidates thus used the net to broadcast campaign materials as used it to interact with others online.

By the week before the election, the debate discussion group had attracted slightly more than eight hundred postings. But fully one-third of those postings were made by only four individuals—none of whom were candidates. The Ottawa Centre riding, the district that encompasses much of the city, had seven candidates, but only one message was posted by one of the candidates from the three major parties. One independent candidate posted ten messages. In the rural riding closest to my home in New York, the Prescott and Russell district, one candidate was a registered user and none of the candidates participated in the online debate.

When one candidate's representative asked for support for the candidate and alerted the forum to that candidate's information section, two messages immediately attacked him for requesting support without backing up his statements. One person derided the candidate for not posting to the list himself. In response to these assaults, a third person, claiming to be not a supporter of the candidate but someone interested in political debate, challenged the attackers to ask the candidate a question; after all, wasn't that the point of this forum? the writer asked. Eventually the candidate was asked a direct question. That question was attacked for its grammar by someone else. The candidate never responded.

The quality of the debate, like the participation, was spotty. My admittedly unsystematic study (I read about forty random posts) revealed a relatively civil if somewhat unenlightening discussion. There was some taunting, a little name-calling, and some reasoned argument. Most of what I saw looked like the statement and restatement of entrenched positions. As in most discussion groups I've seen, the typical post consisted of cascades of snippets of previous messages followed by the writer's response. Most comments seemed to cast the previous statements as wrong thinking. My favorite consisted of a long excerpt from a previous post followed only by "yadda yadda yadda" and the writer's name and address. Overall, my reading provided a glimpse of the candidates' polar positions on general issues (such as taxes, welfare, economy, health care) but did not uncover detailed, substantive positions.

Although the project showed potential, local participation was modest. During the week of May 20 to 27 (the election was held June 8), the menu for the Ontario Provincial Election Project was accessed 1,084 times by 624 unique registered users (2 percent of whom were NCF administrators). This is an average of 1.7 times per person. Of all of the items on the menu, the two most often chosen were item 2, Debate (chosen 287 times), and item 5, Local Ridings and Candidates (chosen 253 times). Given that there were hundreds of thousands of eligible voters in Eastern Ontario and more than forty thousand registered users of the NCF, the impact of the net on the election was probably small.

This example begins to identify the larger issue. The communitarian ideal is for all this connectivity to increase participation in community life. This ideal seems to be manifest in, for example, the amount and range of community-based information provided by some of the large community nets. Just telnet to the NCF, log on as a guest, and begin to explore the choices under the headings Community Associations or Special Interest Groups. The number of options, the amount of information, and the number of opportunities to discuss are staggering. But do members actually use this information? Do they take part in community-based discussions? What do most of them actually do when connected to community nets?

# 11 Reality versus the Communitarian Ideal

If people don't want what community nets are supposed to give them, then a lot of effort will be wasted, says Steve Cisler:

> The premise of most community systems is that the participants want to get local information—job listings, sports scores, community calendars, etc.—and to exchange mail and participate in discussions with fellow citizens. To support access to local information, community network organizers have persuaded government offices, hospitals, and local organizations to input their data and maintain it. But many users are primarily interested in getting out of town, that is, using resources located around the world via the Internet. If the statistics show little local use because the subscribers are spending time elsewhere and not in the local files and discussion areas, the local agencies and businesses may cease to maintain the data files, and the community system will be like a dying mid-town shopping center where the tenants drift away to the suburbs.[1]

Snatches of data support Cisler's concern: members of some community nets are accessing worldwide resources, primarily for entertainment, far more than they are accessing local information.

In a survey for the Blacksburg Electronic Village, Rebekah Bromley found that members use the system primarily for three general purposes: to communicate via e-mail with friends and families, to obtain information related to recreation and entertainment, and to connect to discussion forums related to information and entertainment.[2] Specifically, the survey showed, first, that 97 percent of subscribers use e-mail: 65.2 percent access it at least once a day, and 30 percent check for messages more than once a day. Most subscribers—85 percent—use e-mail to chat with family or friends, whereas 48 percent use it for recreational purposes, 43 percent for professional or academic purposes, and 39 percent for business. Second, the survey showed that 80 percent of subscribers use the network to access such information as databases, library catalogs, and remote computers. Twenty-five percent do this every day or every other day, and 23 percent do this about once a week. "The most frequent use of the network for databases and other sources of information," notes Bromley, "was for recreational purposes (60 percent), such as hobbies and special interest." Finally, 62 percent of subscribers use electronic bulletin boards or newsgroups, and 53 percent access them for recreational purposes, rather than for professional, academic, or business-related purposes.

Because users connect to BEV through gopher and World Wide Web systems that allow wide access to the Internet, I think it is safe to assume that much of this communication and information retrieval takes place via the Internet. Whether to enable users to connect to the Internet is a key decision for any community network. Some offer no such connections. Many others do offer external connectivity, with restrictions. The National Capital FreeNet falls into this category. Because of its operating philosophy and concerns about cost, it restricts its members from connecting via telnet to remote systems on the Internet. The NFC also limits Internet Relay Chat to within its own system, and it does not allow its members to connect to certain MUDs or to transfer files via ftp to and from points on the Internet. These constraints, however, are rather ineffective.

In two papers presented to the NCF Board of Directors, NCF Vice

President Andrew Patrick describes the net's difficulty in enforcing its restrictions on telnet functions.[3] Despite the policy, several "doorways" in the system provide fairly easy and broad access to the Internet. One analysis of system logs, notes Patrick, indicated that NCF users connected to more than thirty-four hundred remote host systems in an eighteen-month period. His list of the twenty most popular telnet destinations from one week in the fall of 1994 includes the following:

Moosehead SLED MUD
Anonymous IRC clients
SchoolNet MOO
Armageddon MUD
MadROM MUD
Finland FreeNet
Channel 1 MUD

"Moreover," he predicted, "as we install the even-more-popular www service (and lynx browser), the number of telnet destinations reachable from our system will grow significantly once again." He was correct. The World Wide Web became quite popular. The bottom line, said Patrick, is that telnet connections are really not restricted.

Patrick described several alternative policies, which either further restricted telnet possibilities or enlarged them. He argued for opening the system up to the world: "It is important to recognize that services like telnet, Gopher, and www provide valuable and important information that is often just as useful to the community as that provided by local organizations. Many information providers on the FreeNet, for example, now include pointers to related information available on the Internet. . . . it may be time to recognize that the NCF is part of a much larger whole, and that Internet services are an important part of the community project."[4] Although Patrick's position is certainly arguable, it may run counter to the primary intent of community nets. Recall Tom Grundner's vision: the success of community nets can be measured not by the amount of international connectivity they provide but only by the intensity of their local connectivity. The question then becomes, Can a community net bring the community to the world and bring the community closer together at the same time?

A glimpse at the NCF's usage data suggests an answer: maybe. Look, for example, at the following list showing how many users chose each option on the main menu during one week:[5]

National Capital FreeNet—Main Menu
Number of Different Users Who Chose Each Menu Item
5/28/95 to 6/4/95

| menu item: | no. of users: |
|---|---|
| 1. About the National Capital FreeNet | 398 |
| 2. Administration | 1,020 |
| 3. Post Office | 6,504 |
| 4. Public Discussion | 1,828 |
| 5. Social Services, Health, & Environment Centre | 409 |
| 6. Community Associations | 411 |
| 7. The Government Centre | 603 |
| 8. Science, Engineering and Technology Centre | 291 |
| 9. Schools, Colleges and Universities | 606 |
| 10. The Newsstand | 1,460 |
| 11. Libraries | 1,269 |
| 12. Special Interest Groups | 2,098 |
| 13. Communications Centre | 3,769 |
| 14. Professional Associations | 401 |
| 15. Help Desk | 455 |
| 16. Menu principal français | 138 |
| 17. Ontario Provincial Election Project | 606 |

Most users chose either e-mail (Post Office) or newsgroup discussions (Communications Centre). If you look at any other week's data, you will see a similar distribution. Most people use the net to communicate with other people (whether those other people are locals is hard to tell). Far fewer members seem interested in items 5, 6, and 7. Although those menu choices do lead to newsgroup discussions, they also point to much information about the local community. Compare, for example, the limited public participation in the Ontario Provincial Election Project, described earlier, to the number of members choosing item 6, Community Associations. This item, which comprises about twenty associations, attracted fewer members than the Election Project, which is essentially a single-item choice. Still, if most of that person-to-person communication takes place within the community, then the NCF is fulfilling one of its prime missions.

Below is a list of the twenty most frequently chosen menu items during the week from May 28 to June 4, 1995. The list identifies choices that lead to online services and choices that lead to asynchronous discussion groups (newsgroups), and also includes other "productive" choices—those identified by the NCF as "menu choices that result in something other than opening another menu."

NCF—Twenty Most Frequently Chosen Menu Items
5/28/95 to 6/4/95

| choice: | no. of users: |
|---|---|
| 1. *Service:* FreePort-menu-system | 13,553 |
| 2. *Service:* mail—mail reader | 9,550 |
| 3. The Post Office—Check Your Mail | 5,580 |
| 4. *Service:* mail send | 4,185 |
| 5. The Post Office—Send Mail | 3,418 |
| 6. Communications Center—Read your favorite newsgroups | 3,136 |
| 7. Communications Center—Read all NCF and USENET groups | 1,863 |
| 8. *Service:* lynx-web-browser | 1,778 |
| 9. IRC Public Chat—Enter the chat system | 1,506 |
| 10. Libraries—Gopher Electronic Library Service | 1,396 |
| 11. *Newsgroup:* ncf.announce | 1,346 |
| 12. *Service:* userInfo—get-from-name | 1,318 |
| 13. *Service:* who | 1,262 |
| 14. Manipulate Files in your work dir.—List your files | 1,232 |
| 15. *Service:* time-remaining | 1,231 |
| 16. *Newsgroup:* ott.forsale | 1,195 |
| 17. Communications Center—Read newsgroup of your choice | 1,000 |

| | |
|---|---:|
| 18. Manipulate Files in your work dir. — Delete some files | 902 |
| 19. Manipulate Files in your work dir. — Delete file by name | 870 |
| 20. *Service:* userInfo — get-from-ID | 845 |

E-mail is king. Items 2, 3, 4, and 5 all involve the mail system (item 1 represents the system's main menu). But again, whether that e-mail activity is local cannot be determined. Newsgroups are second in popularity, and two local or regional newsgroups make the top twenty: ncf.announce is a local list for general announcements, and ott.forsale is a regional online flea market. (In other weeks another regional newsgroup, ott.jobs, ranked quite highly.) Use of the two userInfo services also indicates local communication: many members are searching for information about other members. In contrast, the popularity of the World Wide Web (item 8) and gopher (item 10) reveals that many users are moving out to the Internet.

A look at only the most popular newsgroups provides a clearer breakdown of local and global choices:

NCF — Fifteen Most Popular Newsgroups
5/28/95 to 6/4/95

| newsgroup: | no. of users: | rank in top 100 menu choices: |
|---|---|---|
| 1. ncf.announce | 1,346 | 11 |
| 2. ott.forsale | 1,195 | 16 |
| 3. ott.jobs | 797 | 21 |
| 4. ncf.general | 681 | 25 |
| 5. alt.sex | 509 | 35 |
| 6. ncf.admin | 450 | 38 |
| 7. alt.sex.voyeurism | 401 | 40 |
| 8. alt.sex.exhibitionism | 374 | 43 |
| 9. ncf.government.ont-elec.general | 362 | 44 |
| 10. ott.events | 299 | 50 |
| 11. ott.singles | 275 | 52 |
| 12. ncf.sigs.irc | 251 | 59 |
| 13. alt.sex.breast | 245 | 61 |

| 14. ncf.sigs.games.computer-games | 245 | 62 |
| 15. alt.sex.stories.d | 239 | 65 |

Of the thousands of newsgroups, these fifteen are consistently among the most frequently chosen, and they indicate a nice meld of practical local or regional discussions and global erotica. People want money (ott.jobs), stuff (ott.forsale), love (ott.singles), and safe sex (the alt.sex groups), with a little politics and game-playing thrown in. These newsgroups and the following special interest groups illustrate the range of entertainment and interpersonal support that members seek. (Note that the first two special interest groups are also included in the previous list of newsgroups, as items 12 and 14, respectively.)

NCF—Fifteen Most Popular Special Interest Newsgroups
5/28/95 to 6/4/95

| special interest group: | no. of users: | rank in top 100 menu choices: |
| --- | --- | --- |
| 1. Internet Relay Chat SIG | 251 | 59 |
| 2. Computer Games SIG | 245 | 62 |
| 3. Gay, Lesbian, Bisexual SIG | 192 | 85 |
| 4. ncf.sigs.futurist.x-files | 184 | 88 |
| 5. Comedy SIG | 184 | 92 |
| 6. PC Technical Support SIG | 182 | 94 |
| 7. ncf.sigs.business.gas-prices | 159 | — |
| 8. Stock and Commodity Trading SIG | 156 | — |
| 9. Macintosh User SIG | 156 | — |
| 10. Star Trek SIG | 151 | — |
| 11. Home Based Business SIG | 147 | — |
| 12. Dog Lovers SIG | 144 | — |
| 13. DOS Windows SIG | 141 | — |
| 14. NCF 30-something SIG | 136 | — |
| 15. Radio and Television SIG | 130 | — |

Most of these groups probably consist of locals communicating with each other.

Overall, these data are consistent with the results of a study that recorded NCF members' uses of the "go" command, which enables users to jump to other points in the system. Although some local menu items (such as the Star Trek discussion group) were accessed often, the most frequently chosen items from the NCF menus were ones that led to sources external to the NCF. The authors of the study, noting that services that promote interpersonal communication are very popular, conclude that "these preliminary results indicate that communications, particularly communications that may be directed outside the NCF, are a very important feature of this FreeNet."[6]

Community nets are indeed making it possible for some individuals and some organizations to connect to other people in their geophysical communities. But for some nets this phenomenon represents only a part—probably the lesser part—of their actual functions. There is reason for hope: the movement is in its infancy. Community nets are expanding from zero to thousands of users in less than a year. But there is also reason for concern. One population center may host both a thriving community net and numerous commercial Internet providers, most of which are primarily selling global access. And technological advances may soon bring us e-mail and other communication services through our coaxial cable television wires or through our satellite television receivers. Most of these services will probably offer seductive national and global programming. (Just look at systems like the popular PrimeStar service: you get a compact satellite dish and an eye-glazing number of television channels—but you don't get any local stations.) We must realize, as Cisler notes, that the raison d'être of community nets may be subsumed by the commercial balance sheet: "Many of the services provided on community systems are valuable to the community as a whole, but they may not make much money. Commercial services seeking a healthy return on their investment may avoid marginally profitable services."[7] If community nets attract most of their support because of their Internet connections, they may eventually lose in two ways: first, their communitarian function may wither if much of their resources are used to fund Internet services, and second, they may eventually lose members to companies that provide cheap, technologically advanced, and ubiquitous commercial connectivity.

If community nets become primarily on-ramps to the information highway, the only community-enhancing function left for them to provide is free or cheap connectivity. Community nets would, in effect, train the masses for the commercial net. Recall from the NCF's operating philosophy that one of the net's functions is to "prepare the community for full and broadly based participation in rapidly changing communication environments." Such preparation is under way everywhere. When I discussed the work being done to establish a local community net where I live, I touted the role of our small, independent telephone company. In explaining how it will benefit from the North Country Network, the Nicholville Telephone Company lists "Internet awareness" first: "It would be a very expensive marketing project to bring an awareness of the Internet and its benefits to the approximately 45,000 people in Nicholville Telephone Company's geographic marketing area. The partnership meetings and the activities associated [with the development of the net] have and will do more for educating the general public about the availability and benefits of the Internet than could have been easily accomplished by one company's efforts." [8] Community nets thus will make better global net consumers out of the currently unconnected. That is a reasonable trade-off, if they can introduce participative communication technologies to the unskilled and disadvantaged.

# 12 "Today's Next Big Something"

The goal of community nets—to find a technological fix for our dissolving communities—is not new. Yet because we are increasingly afflicted with that particularly postmodern disability, acontextuality, we tend to forget failed dreams. We insist on seeing today's products, technologies, and social enterprises as unique. So in order to gauge the extent to which community nets can achieve their social goals, it is useful to step back from the current and future uses of the technology and examine a slice of the recent past.

Naming themselves Communitarians with a capital *C,* Amitai Etzioni and a group of "ethicists, social philosophers, and social scientists" in January 1991 published the first statement of their Communitarian Movement, entitled *The Responsive Community: Rights and Responsibilities.* Two years later Etzioni published *The Spirit of Community,* a book that begins with a declaration of Communitarianism. "We hold these truths as Communitarians," the declaration concludes, "as people committed to creating a new moral, social, and public order based on restored communities, without allowing puritanism or oppression." Etzioni refers only in passing to communication technologies, in order to indicate how they may strengthen the "communitarian nexus" that the movement champions. "Postmodern technology," he says for example, "decentralizes labor and capital and can help people work at or near home thereby enabling people more opportunities to put down roots."[1] On the whole, however, Etzioni does not explore how communication technologies can be used to shape the Communitarian revolution.

But twenty years earlier, on July 14, 1971, Etzioni, then the director of the Center for Policy Research, announced that the center had received a grant for $124,300 from the National Science Foundation to study the roles that cable television (CATV) could play in communication and decision-making at the community level. Etzioni was quoted at the time as saying that the study will investigate whether CATV could be the way to "provide neighborhoods with their own TV networks which could be used for community dialogues with elected officials," for community-wide citizen polling, and for communication among members within a community.[2]

Just as Etzioni's concerns echo over two decades, so too do statements regarding the relation of communication technologies to Canadian community development. In a March 29, 1995, report to the Canadian Radio-Television Commission (CRTC) in preparation for CRTC hearings on "information highway convergence," Garth Graham of Telecommunities Canada wrote: "A community network is electronic public space where ordinary people can meet and converse about common concerns. . . . it's an electronic commons shared by all, not a cyberspace shopping mall. . . . Almost anyone can take hold of interactive computer mediated and networked communications and use it to participate significantly in community life and social development. The direct socio-economic impact of a Community Network is that it makes human institutions human again." Twenty-four years earlier, in a 1971 report in preparation for hearings on the integration of cable television into the Canadian broadcasting

system, the CRTC stated that public access television can "enrich community life by fostering communication among individuals and community
groups. . . . Local programmes should be based on access and freedom
from the restraint of program schedules. . . . It is now possible to conclude that community programming is a practical possibility and that
it has considerable potential for further development and refinement as
part of the Canadian broadcasting system."[3]

Other pronouncements from the early 1970s offered similar visions.
A 1971 report from Princeton's Center for Analysis of Public Issues expressed the hope that the technology would enhance civic enterprises:
"Free access public TV channels have the potential to revolutionize the
communication patterns of service organizations, consumer groups, and
political parties, and could provide an entirely new forum for neighborhood dialogue and artistic expression." A speech given on June 26, 1971,
by the FCC Commissioner to the Urban CATV Workshop in Washington, D.C., in which the commissioner read from "An Evening's Journey
to Conway, Massachusetts," by Archibald MacLeish, conveyed the hope
that the technology would help reverse urban decay: " 'I'll tell you what's
a town. It's a meeting of minds. And how do you get a meeting of minds?
Meeting of men. And how do men meet? By the ways. . . . Roads! That's
what a town is—men going back and forth on their occasions. Passing
each other. And not always passing—pausing sometimes—speaking—
palaver. That's what a town is. . . a meeting—a meeting of men—minds.'
So it is with cable's potential. Perhaps as never before has technology
given us the opportunity for a new 'meeting of men, meeting of minds'
in our unworkable inner cities." The 1972 book *Cable Television: A Guide
for Citizen Action* reveals the hope that the home will become the center
of civic exchange because cable television will "make it possible to find
out what's going on in your town, your neighborhood—even your block.
Cable TV can provide *local* information the way the newspaper gives
local news. But it can do *more* than the local paper. Cable Television can
make it possible for your community organization to conduct a meet-

ing of all the people in your neighborhood, without any of them having to leave their homes." Another 1972 book, this one with the prescient title and subtitle *The Wired Nation, Cable TV: The Electronic Communications Highway,* articulates the hope that the new technology will wrest control from a centralized commercial elite: "Together, then, the elimination of channel scarcity and the sharp reduction of broadcasting cost, can break the hold on the nation's television fare now exercised by a small commercial oligarchy. Television can become far more flexible, far more democratic, far more diversified in content, and far more responsive to the full range of pressing needs in today's cities, neighborhoods, towns, and communities."[4]

A number of pronouncements like these were cited in Gilbert Gillespie's 1975 study of public access television, in which Gillespie himself makes some sweeping pronouncements:

Civic bodies must become thoroughly involved with the consideration of design and control of a wired city. For the first time there is an obligation to involve both individual private citizens of the most humble stature and community communications committees in planning the design and future control of an all-pervasive and revolutionary fact of city life. There is now an obligation on the part of the major city governments of Canada and the United States to maintain and invite a defined share of access for individuals and citizens' groups to the proliferating channels of cabled communication in the city. The city fathers must now nurture and eventually react, if they are not already doing so, to many new decentralized sources of local propaganda.[5]

Clearly, people were saying the same thing in 1971 about public access television that they are saying now about the net. And they said similar things in the past about other technologies, also. "This isn't the first time a new medium has come along, promising to radically transform the way we relate to one another," says Todd Lappin. Calling the Internet Today's Next Big Something, Lappin illustrates how current claims about the net are similar to those made about radio in the years before it was transformed into the commercial broadcast industry. At that time, he explains, radio was a grassroots, many-to-many medium whose adherents disdained any kind of commercialization; rather, they saw it primarily as a means to link citizens.[6]

Many advocates of public access television in its early days considered that movement in the same way. Although the first CATV systems began operating in the United States in the late 1940s, it wasn't until the early 1970s that the technology began to expand rapidly. By 1971 twenty-five hundred cable systems across the country served 5.5 million subscribers.[7] But more important, 1971 was a key year for that populist offspring of cable television, public access television.

On July 1, 1971, a seventeen-year-old boy confined to his home following a series of operations to remove a brain tumor participated in a history class via interactive cable television. Jeff Hubert, sitting in front of a video camera and aided by a special keyboard, watched his teacher on his home television while she, in turn, watched him on her monitor in an experimental studio-classroom four miles away. The experiment was the initial step in a plan by TeleCable Corporation of Norfolk, Virginia, to develop interactive cable television to serve communities. The company's president, Rex A. Bradley, commented, "We feel that the addition of a two-way capability means the entire community has the opportunity of acting on the cable. A politician can propose a new idea and ask the viewers what they think of it. The people can push the $Y$ button if they say yes or push the $N$ button for no." On that same day TeleCable previewed the commercial capabilities of interactivity by demonstrating, as one article put it, "how a wife could shop from home" by viewing a commercial for laundry detergent staged at a local Sears store and following the instructions "on what buttons to push" to order the detergent.[8] Observers and practitioners predicted that the new technology would provide up to fifty channels of recorded music; offer specialized news services in which a subscriber could enter codes to receive news about areas of interest; enable subscribers to "dial up" pre-recorded pictures and such textual information as consumer organizations' ratings of products; provide "electronic mail delivery that would, in effect, print out telegrams at home or office"; and make the television keyboard and screen a computer terminal "by which the subscriber could dial up specific reference information and perform mathematical calculations" and have the answers printed on the screen.[9]

Also on the same day that Jeff Hubert had his first tele-history lesson, a group of New Yorkers celebrated the beginnings of free access to television technology with a day-long Washington Heights block party, highlighted by a public argument between an executive for a cable television company and an advocate for public rights.[10] The argument concerned two Manhattan cable television stations, TelePrompTer and Sterling Manhattan Cable Television, which on that same day had become publicly available—groups or individuals could apply for free airtime on a first-come, first-served basis. "Supporters have hailed the program as the first genuine 'Town Meeting of the Air,'" said George Gent in a *New York Times* article, "and a major step toward the political philosopher's dream of participatory democracy."[11] A TelePrompTer brochure defined the station as a "whole new concept in television." It is "TV by the people, for the people," permitting "any groups or individuals of any belief, purpose, or persuasion, to demonstrate their talents." If people did not have the equipment or expertise to take this opportunity, the station would provide free studio space, at least one television camera, and a director. It would also provide portable equipment "to cover events in the community, like block parties, park openings and church functions."[12]

That was the good news. The bad news was that because of uncertainties about liability, the cable companies had established rules controlling citizens' uses of the new medium—controls that at least one public advocate said were too restrictive. The rules determined how long one person or group could use the channels at any one time and set a minimum age for users (no one under eighteen could use the channels without the presence of an adult). More important, the rules stipulated that the company had the right to choose not to air any program developed by any citizens. At the Washington Heights block party, this issue

> sparked the argument between Irving B. Kahn, TelePrompTer's chairman, and Theodora Sklover, executive director of Open Channel, a nonprofit organization to aid community groups in planning for the use of public channels. . . .
> Mr. Kahn approached the microphone in the middle of a roped-off St. Nicholas Avenue between 178th and 179th Streets, and edged aside the Open Channel head who was acting as mistress of ceremonies. He accused her of distortion and unfairness.
> He asked her to retract her statements. She said "I can't retract."
> At one point, Mr. Kahn asked her: "If they can come on what more access do they want?"[13]

Sklover and other neighborhood activists had also complained that the restrictive rules had been written without a public hearing, effectively shutting out the views of prospective users of the new medium. This argument revealed that the issue at stake is not merely one of access but also one of expression. That public fight foreshadowed similar problems that face public access advocates today, advocates for both citizen's cable television and community computer networks.

Many observers of the new technology in the early 1970s argued, however, that through CATV local information generated by civic groups and citizens could be distributed more widely than it could through other means. The civic and political implications of this feature were discussed at length in the 1973 book *Talking Back: Citizen Feedback and Cable Technology*, edited by Ithiel de Sola Pool. In his chapter, "Citizen Feedback in Political Philosophy," de Sola Pool argued that public access television would not necessarily increase the effectiveness of political decision-making because large-scale political feedback systems were too open to abuses: the technology could be used to fuel demagogic hysteria among political activists or to surveil and manipulate citizens. Instead, de

Sola Pool maintained, the technology would best be used to enable citizens within a locale to communicate with each other, thereby creating a bottom-up political force: "Perhaps, then, the most important way to think about citizen feedback in the CATV era is not as a device that the President or the Congress will find easy to use, nor as a device that will give citizens much more voice in those top-level decisions, but rather as a device that will promote grass-roots interactions among citizens with special interests."[14] Yet some public access activists of the time sought an even more powerful function for the new technology: that of transforming individuals and communities.

The relation between communication technology and the social transformations of community that were the goal of the public access movement underlies Gillespie's 1975 study of public access television in Canada and the United States. The activists assumed that the act of making or viewing electronic images of one's self objectified the self, making it a separate entity that could then be examined, better understood, and possibly altered. Likewise, objectifying community actions could lead to the same kinds of transformations. Gillespie traces this theory back to the mid-1960s in the work of filmmakers associated with the National Film Board of Canada (NFB) who were making documentaries about distressed social and economic conditions in different parts of the country. One film, *Septembre 5 at Saint-Henri,* provoked a negative reaction in the Montreal community where it was filmed. Some of the residents depicted in the film were ridiculed, and some were so ashamed by the portrayal of themselves and their lives that they withdrew their children from school. One of the NFB filmmakers was shaken by this experience, and when he had an opportunity to make a film in another part of Quebec, he and his colleagues invited the subjects of the film to participate in its making. By encouraging the participants to view the ongoing production, they expected to learn from the subjects how best to proceed with the film. But the makers of the film discovered that, for the participants, the process of viewing themselves on film began to influence their behavior. These screenings, noted the filmmaker, had the effect of "collective therapy" for the subjects of the film.[15]

Filmmakers of the NFB undertook a similar experiment in 1967 on Fogo Island, a remote spot located ten miles off the northeast coast of Newfoundland that had undergone tremendous economic and social decline as the local fisheries had become depleted. The island was home to ten settlements, each isolated and mistrustful of the others. Some islanders grew up without ever visiting another settlement. The NFB chose this site because it seemed to be a microcosm of poor, rural, isolated areas in Maritime Canada. The main purpose of the filmmaking process was to show the Fogo Islanders to themselves. "The people," said Sandra Gwyn in a report on the progress of this experiment, "were trapped in a cycle of isolation and poverty from which they lacked the knowledge and confidence to escape." In order to begin the process of helping them help themselves, the NFB brought in the equipment that enabled them to tell their stories to others: "The final result was 28 short films adding up to a total of 6 hours, each centered round a personality or an event rather than an issue; each expressing an aspect of life on Fogo Island. . . . Through looking at each other and themselves, Fogo Islanders began to recognize the commonality of their problems as important, they began to become conscious of their identity as Fogo Islanders; they discovered that preserving the Fogo environment mattered to nearly all of them."[16]

Within a few years, video had become more available, and programs in the spirit of the Fogo Island effort were channeled through CATV media. A technological breakthrough occurred in 1968 when Sony Corporation began to sell the video portapak that enabled mobile taping. Community members could be given video equipment and training in order to develop fledgling electronic information infrastructures. In 1972, the CRTC conducted a study of public access cable television in Canada and found that there were 361 Canadian CATV systems, of which approximately one-third were offering public access. As part of the study administrators of 82 systems were surveyed to determine their attitudes toward public-generated, "local origination" programming. The vast majority were positive. When asked if local origination is best when community individuals and groups participate, 63 percent agreed or strongly agreed. And 72 percent agreed or strongly agreed when asked if a major concern of their station was to "facilitate the emergence of new forms of community expression." [17]

By the early 1970s, notes Gwyn, the "Fogo Process had been adapted for use all over North America; the NFB's . . . units seeded community communications groups all over the country, who produced local programs for CATV systems." Much of this work was done in New York City, because of the concentration of technology and expertise available there. As Michael Shamberg described in the 1971 book *Guerrilla Television*, one group, People's Video Theater, developed a Fogo-like process to mediate conflict. They might videotape one side of a conflict and show it to the other, and vice versa; then they would tape the opponents' responses to the first tapes and show the second tapes to both groups together. "They've developed techniques of video mediation which essentially cool people out by taping them and letting them relate to one another through the medium," said Shamberg.[18] Another group, Videofreex, went on the road with the portable technology, teaching the locals of various communities how to create their own video and spread the faith in electronic imaging. Shamberg commented,

> You go out to communities and do videotaping; they pay you to come. Ideally you plug-in to existing hardware and show people how to use what they already have.
>
> In smaller communities especially, people are hungry for novelty. With a media bus you can entertain them while you're there, and then turn them on enough for them to want to set up their own media system after you've left.[19]

The heir to much of this work is the guerrilla television that arose during the 1980s through organizations like Paper Tiger TV. As bulky and expensive Sony portapaks evolved into cheap camcorders, the costs of developing video dropped, allowing activists to produce projects that the television industry had missed or ignored. "A case in point is *Drawing the Line at Pittston* (1990)," notes Paper Tiger participant Jesse Drew, "a tape made in collaboration between Paper Tiger members and striking mine workers in Virginia's Pittston mines. . . . When I showed this tape at a large meeting of union members, the response was overwhelming. For many, it was the first time they had ever seen television depict the struggle of working people in a positive and realistic light." [20]

Although now camcorders are ubiquitous and organizations like Paper Tiger have the reach of satellite transmission (via Deep Dish Television), any large-scale effects of these efforts are hard to see. But in 1975 Gillespie seemed sure that public access to television technologies could indeed transform society: "And so the evidence is in. It is as a hammer, not as a mirror, that PACT [public access cable television] can create social change, can stimulate social animation. It can and should be used as a hammer. It is, therefore, to some degree, a cybernetic process of synthetic reproduction of men's intelligence, shared intelligence with the occasional objective of reaching a consensus." [21] Gillespie's argument still resonates twenty years later in beatific visions of the hive mind of the net. But with public access television, these moments of shared intelligence were discrete and few in number. Although public access television has had significant impact in focused situations, it has not fared as well across the broad spectrum of the populace that the net, its proponents claim, is destined to attract.

At the height of optimism about public access television in the early 1970s, the medium had little reach, even within communities. In 1974 the Institute for Communications Research at the University of Indiana attempted to measure the audience for public access television by conducting a case study of a community that seemed to be a prime candidate to embrace the new technological opportunity. This community was Columbus, Indiana, a town that had at the time about thirty-five hundred CATV subscribers and no local stations. Columbus promoted itself as the "Athens of the Prairie." It was known for its diverse architecture and its support for the arts, and its public access television facility was active, well staffed, and well equipped.

To assess the size of the audience, the institute studied a typical week of television viewing by collecting viewer diaries from 6 percent of the subscribers, encompassing 643 people in 200 households. The diaries revealed that only 20 people (about 3 percent) had watched any public access television at all, and 19 of them had watched less than one hour's worth during the week. The remaining person, a fifty-year-old woman, was a fan of public access and had spent 40 percent of her viewing time watching it. The 19 others watched an average of 4.7 hours of television a day, while the public access fan watched 6 hours a day. In a review of this study Rudy Bretz puts these numbers in perspective.

> The important, and embarrassing, question is—how many people were watching any given public access program, at any given moment? The public access channel had been, as usual, programming something for at least 57 hours during the test week. . . . Eighty percent of the programming was not reported to have been viewed at all, by *any* member of the sample group.
>
> The total public access viewing time, for all persons in the sample, was 2/10 of one percent of the total of all television viewing for the week. Because this is about the rate of error for such a study, the true figure might be less, or it might be as much as twice as great. Even so, it would be very small.[22]

Public access programs that aired during peak viewing time, when 2,345 subscribers (67 percent) were watching television, probably attracted no more than nine or ten households. Bretz speculated that even if two people at each household had been watching (which was the average for the study), thus raising the total to eighteen or twenty viewers, each program would have had about the same number of viewers as there were staff and volunteers working at the public access station.

In response to these numbers, Columbus Video Access Center decided to promote its programs by listing them in the local newspaper and by offering a local news show—the only local news in Columbus. A telephone poll conducted during the news program indicated that about 3.8 percent of the televisions in use were tuned to the public access news show. Other estimates of nationwide public access viewing that Bretz analyzed indicated that 50 to 190 people per 100,000 potential viewers might be watching public access television during a prime viewing time.[23]

While commercial cable television companies have evolved into media giants, the populace has never taken advantage of public access television. In the early 1970s my counterparts—writers touting community development through the latest technology—warned that the window of opportunity for public access television could eventually close. Monroe E. Price and John Wicklein stated,

> Cable has gained so much momentum already that there is little chance it will be stopped. But if the Wired Nation does emerge, for whom will it be wired? If the public does nothing, the answer to that is easy: The nation will be wired primarily for the benefit of private entrepreneurs. Cable will then be much like broadcast television and radio before it. Programming will be restricted to mass-appeal entertainment, superficial reporting of news, and minimal discussion of public affairs. Cable subscribers will be sold to advertisers at so much a thousand, as the over-the-air audience are sold to them today. Community service and public access to the systems will be given lip service only, as they are in most commercial television and radio broadcasting. The opportunity for a revolution in communication through cable television will be lost.[24]

I am tempted to say that this opportunity has already been lost, but evidence suggests that community development aided by public access television is still quite active in some locales. In these areas the stations are attempting to accomplish the same kinds of things that community computer networks are attempting to do.

When the city of Minneapolis signed a new city cable franchise agreement in 1983, it created the independent, nonprofit Minneapolis Telecommunications Network. This system manages a number of local public channels and "provides coverage of neighborhood and community events, assists the City and the Minneapolis Public Schools in the use of communications technology, and provides access to communications technology for the residents and communities of Minneapolis." The system has more than five hundred volunteer producers and workers and comprises one hundred nonprofit organizations, and its productions have taken place "in every neighborhood and community in Minneapolis."[25]

According to Lauren-Glenn Davitian of Chittenden County Television (CCTV) in Burlington, Vermont, the public access movement has succeeded with stations like MTN and CCTV because these stations have met the needs of their communities for low-budget media, they reach many people within given areas, they put portable equipment in people's hands free of charge, and they provide an opportunity for community members to critique the community. Further, in an environment in which even letters to the editor are edited for content or space, public access television is often the only forum for unedited community-based information — especially if the community does not have a computer network in operation.[26]

Such an agenda addresses one of the primary challenges to any public access station and to any community computer network: how to achieve both community development and a viable form of freedom of speech. Note the emphasis of both in MTN's description of its mission:

> Community cable television is a media alternative to commercial television channels. It provides an organizing tool for those dealing with social concerns. It provides visibility for celebrations directly affecting the local community. It provides services that help the nonprofit sector and the community-at-large save money. It is active participation both behind and in front of the camera. It is the opportunity for people to learn about each other's needs, concerns and dreams. For a diversity of voices and opinions to be heard and/or seen, MTN provides 18 channels for those wishing to cablecast non-commercial public programming. All programs produced and/or presented by a resident of Minneapolis are shown on MTN's channels free of charge. The content is the responsibility of the producer with no censorship from MTN — this is a freedom of speech few media outlets allow.[27]

The balancing act is to encourage community development while not letting legal but anticommunity expression threaten that development. Davitian emphasizes that although the stations must remain staunch supporters of freedom of expression, the primary agenda of public access television is community development. Overall, these operations "are about community organizing not technology." [28]

But the agendas of public access television stations, like those of community nets, are threatened by economic uncertainty. Despite all the good work being done in both arenas—despite all the constructive, enthusiastic efforts of their members—both movements may wither in the face of fiscal uncertainty and other external threats. And the impact of public access television on community life has been inconclusive, while the commercial television industry has transformed our lives. Likewise, the future impact of community nets on community life may also be small, while global nets will certainly transform our lives again. Our only hope for preserving community is to fight for community-enhancing applications that are designed or chosen on the basis of local needs—even though we can never be sure of the long-term effects of those applications.

In 1975 Kas Kalba considered the future of television. Kalba's vision of future technologies was quite perceptive, if limited. He did not foresee the evolution of computer networks, but he did predict the rise of the electronic community. And he concluded with a warning against complacency and trust in the marketplace, in the government, in technologists.

In short, the challenge of ensuring that the electronic community enhances the quality of life and encourages active and equitable involvement in this new community sphere cannot be overemphasized. Our recent experience with the physical and social environment in which we currently live has hopefully shattered our naivete concerning how a convivial yet complex living environment is brought into being. The naivete of the business firm that claims it is only adding a new product or technology to the marketplace; the naivete of government that formulates its policies in response to short-term political pressures rather than long-range communications priorities; the naivete of the systems planner or engineer who believes that a neatly-drawn blueprint can anticipate the needs of a dynamic, pluralistic society; and the naivete of the citizen who leaves decision-making about the future up to others until that future impinges on his doorstep: these are not adequate postures for the building of a new communications environment.[29]

Just as it is naive to trust the design of the net to the technotopists, it is equally naive to assume that by turning off our televisions and boycotting the net, we can somehow recapture something we've lost. The only long-term option is to work to use the technologies for the local good.

Recall the mother who has never touched a computer, who has never used a fax machine, who may not have a telephone, whose use of broadband communication technology begins and ends with television, who cannot type, and who has few, if any, technical skills. I asked if she and others equally isolated could be reached by the wired communitarians. If there is an affirmative answer to this question I suspect it refers to some small achievements in some unexceptional places by people employing some mundane technologies. Even so, these activities give me hope—hope for the fledgling North Country Network, for the vast NCF, and for community nets in general.

## 13 Fight the

## Good Fight

Downstairs from CCTV, the public access television station in Burlington, Vermont, lies the Old North End Community/Tech Center, which is charged with the mission of "connecting people to resources so that we can build a community in which no one is left out."[1] There, in conjunction with CCTV, you will find public access computer terminals and a schedule of training classes. During the summer of 1995 the first classes in video production and computer skills were offered free to community members in conjunction with Burlington College and the Department of Employment and Training. Students in the computer classes learned the basics: how to use a mouse, how to store and retrieve files, how to print. In addition, Champlain College offered an introductory course in computer networking, which Old North End residents attended.

As stated in its July 1995 newsletter, the Community/Tech Center

- Serves all neighbors (in the Old North End), including the working and unemployed, long-time Vermonters, New Americans, citizens, and refugees.

- Incorporates community institutions, such as a library branch, literacy brigade, after-school program, teen outpost, senior destination, public access media center, and civic computer network.

The center is

- Built upon a well-researched inventory of neighborhood infrastructure, community needs, employment opportunities and long-term trends.
- Built upon affordable, easy-to-use technology for a range of individual and community uses.
- Run by neighborhood trainees with a community-based Steering Committee.
- Anchored by micro-business development, job training, and placement initiatives.
- Supported by local business and institutions, national foundations, and self-supporting initiatives.
- Measured by concrete benchmarks demonstrating impacts on education, employment, and standard of living.[2]

The center is undertaking the kind of efforts needed to begin the process of networking electronically a neighborhood that is not yet technology literate. This is the work of a proto–community net: the mundane, incremental, but essential actions that might help develop that critical mass necessary to make a community net happen.

And if programs like these work, will they just serve up the neighborhood to the global net? Or can they spawn vibrant community networks geared to serve neighbors and the places they live? It is too soon to tell in the Old North End, but I am encouraged by the Community Associations menu of the National Capital FreeNet.

One of the least popular items on the NCF main menu is item 6, Community Associations. If you choose that option, you will find a mixture of information archives and discussion groups that attract only a handful of members on a weekly basis:

Fifteen Most Popular Community Associations
5/28/95 to 6/4/95
community association:                                    no. of users:

1. Arts, Music, and Culture Assoc.     37
2. Community Assoc. Discussion     31
3. Ottawa South Community Assoc.     22
4. Volunteer Centre of Ott-Carleton     21
5. Consumers' Assoc. of Canada     21
6. Ottawa Naturists     17
7. Hosteling Int.—Ontario East     17
8. Conf: Comm. Access to Info. Hwy.     13
9. Citizens for Safe Cycling     12
10. Expo 2005     11
11. Scouts Canada, Nat. Cap. Region     9
12. Canadian Kennel Club     9
13. Manordale-Woodvale/Arl. Woods     8
14. Ottawalk     8
15. Humanist Assoc. of Ottawa     7

It would truly do a disservice to claim on the basis of these figures that the NCF is failing in its mission to serve communities. If you explore one of these items in particular—the Ottawa South Community Association— you will find the beginnings of the kind of community-enhancing environment that community nets are supposed to be creating.

When you choose this option, you see the following menu:

<<< Ottawa South Community Association >>>

1. About the Ottawa South Community Association
2. Organization
3. Questions and Answers >>>
4. Board of Directors

---

h = Help    x = Exit FreeNet    p = previous    u = up    m = main

Items 1, 2, and 4 provide the standard information: names, phone numbers, descriptions of the area and of the purpose of the association. When I checked out the Board of Directors, I learned that only two out of nineteen board members were registered members of NCF. But when I chose

Questions and Answers, I discovered a possible future of the community net movement.

This discussion group was nowhere near as lively as the big ones: the alts and the forsales and the singles. It had accumulated a little more than a hundred posts in about a year. Further, almost half of the comments had been made by one person, the past president of the association, who appeared to be trying to prime the discussion pump. But what was intriguing about this discussion were its tenor and topics. The participants were engaged in intense debates about local traffic problems, the presence of police in the area and their efforts at community policing, property taxes, and Internet access for the community (specifically, the possibility of getting a public terminal and posting printouts in a central location). Some discussions became tense: a number of posts about the possibility of changing local traffic patterns revealed significant disagreements between some residents who live on busy streets and some who live on quiet streets. Yet interspersed with these exchanges were announcements about local events, requests for help or advice on a community project, and some reasonably neighborly chatter. Nearly all the discussion was circumscribed by the physical locality. As I read, I wondered if this forum was the beginning of not a community net but a neighborhood net.

In a proposal to the NCF, Chris Bradshaw offers the Neigh-Net as the next step beyond the Free-Net in the evolution of community nets.[3] Pointing out that Free-Nets can become large, impersonal, and chaotic, Bradshaw describes the less-is-more philosophy that must guide the way to universal access. "Neigh-nets," he says, "are the next natural step in coping with both the sheer growth in participants and in making the service tangible and relevant." A Neigh-Net can provide a number of services, none of which are spectacular but all of which involve the essential functions of daily living. Neigh-Nets, according to Bradshaw, can provide a medium for neighbors to

help people adjust to changes in neighborhood populations by welcoming new neighbors, saying farewell to departing neighbors, and so on,

market, on a neighborhood scale, such organizations and services as baby-
  sitting, home businesses, and real estate information,
support the development of new neighborhood commerce and businesses
  and help already existing businesses better serve the locality,
debate local issues, from taxes to traffic to crime to politics,
publicize neighborhood events,
make bookings, appointments, and connections more easily, and
find others with shared interests within the neighborhood.

Moving these functions online might even help people focus their attention on the places in which they find themselves. "Putting networking to
work at the local scale," Bradshaw notes, "is a way to support local life
and stem the trend toward thinking that what is near-at-hand is, well,
pedestrian."

Bradshaw suggests a design that makes the Neigh-Net an entity separate from a Free-Net but still connected, so that the two kinds of
networks can function efficiently and share services beneficial to both.
Whether it is better to develop a dedicated net on the neighborhood
level, to create neighborhood centers in larger nets (as the NCF does now),
or to achieve some sort of hybrid is something for community net developers to test and neighborhoods to decide. What is important is that
community nets—in their drives to build and to enlist members and
to keep up with the ever-changing technology—keep their focus on the
micro-community.

Such a focus is consistent with the Neighborhood Agenda program as
described by Ed Schwartz, president of the Institute for the Study of Civic
Values in Philadelphia. Schwartz argues that the work to revitalize Philadelphia must start at the elemental level: "The Neighborhood Agenda is
about housing and trash and the environment and crime and schools and
jobs within reach of where we live. It's about grassroots democracy—the
sort of democracy that a great many people talk about on Internet Lists,
whether they have to do with community issues or electronic community networks. It's bringing to life what the framers had in mind when

they proclaimed that 'we the people' were creating a government to 'insure domestic tranquillity' and 'promote the general welfare' and 'secure the blessings of liberty to ourselves and our posterity.' "[4] Neighbors arguing about traffic patterns may be the least glamorous, lowest-tech thing happening on the net right now. It may also be the most important.

In a 1995 report entitled "The Promise and Challenge of a New Communication Age," the Morino Institute lays out ten general guidelines for people interested in beginning the process of shaping nets for their communities.[5] Although these suggestions are constructive and hopeful, I believe that they need to be complemented with some advice about walking the middle way between denial and obsession, between antitechnological hopelessness and electronic ecstasy.

*Decide How You Will Handle the Demands of the Net.* The first three Morino Institute suggestions involve the education of the would-be community net activist and illustrate two interrelated paradoxes that the activist will face.

1. Learn about the new medium. . . . Educate yourself about interactive communications, the effects it is having on society, business, jobs, government and communities—both the good and the bad. Even if you are not a user of this medium now, realize that your children will be, either directly or indirectly. Understand the opportunities it can open for them and the threats it may present. Recognize that they will be at a disadvantage if they are not prepared.

Here is the first paradox: how do we become net-savvy and, more important, how do we raise net-savvy children without becoming immersed in and seduced by life on the net? To deny our children this knowledge and experience is to deny them access to the culture (remember Porush's Law). But to immerse ourselves and our families is potentially to deny our natural environments. The only alternative is to seek out ways electronic virtuality points us toward our geophysical places and cultures. These ways will be few but potentially valuable.

2. Get connected and establish a network presence. . . . Don't simply subscribe to a service, use it. Explore the information sources, but more importantly, communicate. Find someone to contact. . . .

3. Gain the critical skills and literacies. Yes, using interactive communications will take some work. It's not like television where information or entertainment are delivered with the press of a button.

These suggestions lead to the second paradox. The net is a massive time sink. If you spend the time to become skilled, be prepared to ignore some other part of your daily life. (Check out Cliff Stoll's *Silicon Snake Oil* for some examples of how time-consuming becoming connected can be.)[6] Be suspicious of any net enterprise that defines interactivity as simply clicking a mouse from display to display. (Most of what we find on the World Wide Web is merely display.) Television may be only minimally interactive, but following hypermedia links from one presentation to another doesn't rank much higher on a scale of interactivity—especially if those presentations are primarily graphical. If the net is going to transform communications for the better, it will engage us less in viewing displays and more in substantive interactivity—the kind of work in which we actively manipulate symbols. This kind of interactivity, which ultimately may be needed in order to reintegrate people into geographic communities, may take us away from face-to-face interactions.

*Temper the Technotopists; Navigate through the Naysayers.* The next few suggestions of the Morino Institute involve the ways the activist interacts with the people and institutions of a community.

4. Volunteer for a community effort. Discovering the benefits of the new medium is often best done collectively. Volunteer with a school, library, public service group, church or youth center trying to improve its organization or educate others through the use of interactive communications. . . .

5. Change your schools. If we do nothing else to cultivate the benefits of interactive communications, it should be to help more children

use the new medium. It opens for them new horizons, new ways of learning and communicating and can engage those outside the mainstream. Fundamentally, it will be critical to their future success in life and career. . . .

6. Formulate a blueprint for your community. Insist that your local community, your region or state has a strategic blueprint for entering the Communications Age. It should be a long-range plan that considers the needs of the community at large as well as the specific needs of particular groups and members. . . .

7. Demand awareness and change in the government. Government can, and must, be an integral part of the communications revolution. Its role is crucial. Some local, state and federal leaders and agencies are aware of the changes and potential, but many more are not. Too often they are focused on the technology rather than on the individual, social and economic implications.

As the net and its proponents have achieved celebrity status, a community will not be hard-pressed to find technotopists in its midst. Nor will communities find a short supply of naysayers among its leadership—especially when it comes time to spend local monies. What communities need are people who have some technical skills, a willingness to examine how electronic communication technologies can enhance the community, some drive, and a healthy dose of constructive skepticism. Bring doubt to every claim about the net, but be committed to moving forward. Be suspicious of politicians who suddenly see a bright future thanks to technological breakthroughs; indeed, try to find out if your government officials really know about the technologies in question.

Most important, actively investigate how communication technologies are integrated into your schools. Examine the nature of any distance learning programs your schools wish to adopt. Be sure that those programs are designed to engage students in real-time interactions with teachers and peers and not merely to allow them access to displays of information. Further, if administrators and teachers want net connections merely so students can find cool web sites, object strongly. If administrators and teachers want net connections merely so students can point and click their way through hypertext displays, object strongly. If the technologies are not used in part to teach students about their localities, about communicating with others within those localities, about bridging differences among people within localities, object strongly. To induce a student to engage constructively in conversations with students around the globe is wonderful. But if that is the exclusive goal of the school's connectivity, object strongly.

*Establish Ongoing Discussions of Community Standards.* The final suggestions of the Morino Institute sound reasonable but will lead to some of the most difficult decisions community nets will make. How do community members decide what the common good is? What happens when individuals wish to have access capabilities that go beyond what the community net can offer?

8. Use common sense. The potential of interactive communications is great, but every idea about it is not. In pursuing efforts to use, apply and support projects, think practically. . . . Apply the "human relevancy" test and avoid getting trapped in the technology. . . .
9. Insist upon tangible results. This is of ultimate importance. It is easy to become swept up in the excitement and potential of the technology, forgetting that the end goal is positive change. Building a public access network or helping a service group come online should not be the end result. Enhancing community action or helping the group improve its services should be.

Recall that the National Capital FreeNet states publicly that its directors will decide what constitutes abusive and destructive online behavior. Consider the possible threats to free speech that are engendered by public access television. Those who would build a community net will need to establish a process whereby members of the community and the net continually discuss local standards of behavior. If there is an enterprise that can quickly find the intersection of individual rights and the common good, it is a community net.

10. Encourage bold solutions. Think big, but even more importantly, think new! The promise of interactive communications is not one of small change and modest productivity increases. It is one of sweeping transformation and innovation: the chance to bring together dozens or hundreds of people to address a community problem, for children in a small rural town to work together with other children in an inner city, for governments to really reinvent the ways in which they deliver services.

But after you think big, think small. Think incremental. When you think new, don't forget to respect community traditions that do not impinge upon individual and group rights. Keep in mind all the local people who will need a large amount of help before they ever sit down in front of a screen. Think through each move and where that move will lead. Grand visions are important, but healthy action is stepwise and measured. The net inspires sweeping proclamations about revolutions, but the constructive power of the net is demonstrated in small ways in local places.

The other day I saw what I have decided is the saddest television commercial ever. The opening scene shows what looks like three generations of one family seated around a big dining room table while the elders present the family meal. As the camera moves from one person to the next, the narrator, who is one of the family, tells us a little bit about each person, including himself. At first, the ad seems to be showing us just another idyllic, Rockwellian image of the kind of intact family that we vaguely recognize but rarely experience anymore. But what we think we see is quickly undermined when the narrator tells us how bad mom's cooking is. Then he tells us each person's story: every one of the siblings, in-laws, and significant others lives in a different corner of the country, and they don't see each other very often. Near the end we get a glimpse of the narrator's children, who live with their mother thousands of miles away from him. When we finally see the narrator, we realize that while all the rest are busy chatting, laughing, and eating, he is staring benignly at them all, thinking, "Hey, this is my big and wonderful and crazy family; we're not perfect, but we're good." And the viewer can tell that the members of this splintered family really know each other intimately. The easy banter shows us that they are comfortable with each other. We know that this is a loving family because one of the significant others, an outsider, is the slightest bit reserved, while one of the in-laws is clearly as much a part of the family as the blood relations. The unstated premise is that there is a core to this family, there is a center. If asked what the core of this family is, we are supposed to answer, it is love. If asked where the center of this family is, we should answer, it is in their conversation. The advertisement, of course, is for long-distance telephone services. The message is that you don't need to be together to be "friends and family" or to "touch someone." The world—our jobs, our desires—may forever separate us, but we can overcome this distance through the net. That's the lie of this advertisement and the essential message of this age.

One night last winter when I was returning home from a business trip, I had a moment of realization not unlike that of the narrator in this ad. I had finally reached the north country and had been driving through the cold clear darkness across thousands of acres of desolate, frozen fields—fields broken only by darkened treelines and rock walls, dotted with the occasional barn or double-wide trailer—when I crested a hill and saw on the farthest northern horizon the faint glow of Montreal. Even though the hour was late and the temperature below freezing, I pulled off the road and stopped in the heart of one of the most remote stretches in the northeastern United States, and I looked out on the representation in light of one of the world's great cities. The contrast between where I stood—my home territory—and where I gazed—a shimmering, electric specter of a different, seductive place—was vivid. I stayed there in the quiet for a minute or two. Where I stand is where I live, I thought. It's here and now. It's real.

The net, like the glowing city I gazed at, is a seductive electronic specter. Take part in it not to connect to the world but to connect to your city, your town, your neighborhood.

## Civic Organizations

This list is an excerpt, reprinted here with permission, from a 1995 version of the Electronic Frontier Foundation's list of organizations that promote online or offline community development activities. The entire list is available via ftp, gopher, and World Wide Web (see the contact information at the end of the list). Any questions about this list should be directed to the Electronic Frontier Foundation (e-mail ask@eff.org).

### Online Activism Organizations List, Ver. 7.06

Outposts on the Electronic Frontier — International, National, Regional & Local Groups Supporting the Online Community
An ACTION/EFF FAQ by Stanton McCandlish <mech@eff.org>
Updated: May 11, 1995
Archived at: ftp.eff.org, /pub/EFF/Issues/Activism/activ_groups.faq
See also /pub/EFF/Issues/Activism/activ_resource.faq, the Online Activism Resources List.

### Organization Listings

Canada
*Electronic Frontier Canada (EFC)*
EFC was founded in January 1994 to ensure that the principles embodied in the Canadian Char-

ter of Rights and Freedoms are protected as new computing, communications, and information technologies emerge. EFC was co-founded by Dr. Jeffrey Shallit of U. Waterloo and Dr. David Jones of McMaster U. EFC has established an announcements mailing list, efc-announce, and an open mailing list forum, efc-talk.

General: efc@graceland.uwaterloo.ca

Dr. Jeffrey Shallit <shallit@graceland.waterloo.ca>

Dr. David Jones <djones@insight.mcmaster.ca>

Mailing list subscriptions: <listname>-request@insight.mcmaster.ca, message body: "subscribe <listname>" (see above for listnames)

FTP: insight.mcmaster.ca, /pub/efc

Gopher: insight.mcmaster.ca, 1/org/efc

www: http://insight.mcmaster.ca/org/efc/efc.html

Voice: +1 519 888 4804, Dr. Shallit, +1 905 525 9140 x24689, Dr. Jones

Fax: +1 519 885 1208, Dr. Shallit, +1 905 546 9995, Dr. Jones

Snail: 20 Richmond Avenue, Kitchener, Ontario N2G 1Y9 Canada

### Telecommunities Canada (TC)
["Striving to provide support to all community networks throughout Canada." Founded Aug. '94. No other information is handy as of yet.]

General: Andre Laurendeau, esq. <andrel@pubnix.qc.ca>, President

Voice: +1 514 278 1664 (Andre Laurendeau)

Snail: 1030 Beaubien e. #201, Montreal, Canada H2S 1T4

## USA

### Alliance for Community Media (ACM)
The Alliance for Community Media is a national, nonprofit membership organization committed to assuring everyone's access to electronic media. The Alliance accomplishes this by disseminating public information, advancing a positive legislative and regulatory environment, building coalitions, and supporting local organizing. Founded in 1976, the Alliance represents the interests of over 950 public, educational, and governmental ("PEG") access organizations and local origination cable services throughout the country. The Alliance also represents the interests of local religious, community, charitable, and other organizations throughout the country who utilize PEG access channels and facilities to speak to their memberships and their larger communities. [Note: This ACM is not to be confused with the Association for Computing Machinery.]

General: Barry Forbes <alliancecm@aol.com>, Executive Director

FTP: ftp.eff.org, /pub/Groups/AllianceCM/
Gopher: gopher.eff.org, 1/Groups/AllianceCM
www: http://www.eff.org/pub/Groups/AllianceCM/
AOL: AllianceCM <alliancecm@aol.com>
Snail: 666 11th St. NW, Suite 806, Washington, DC 20001-4542 USA
Voice: +1 202 393 2650, +1 202 292 2653

*American Society for Information Science (ASIS)*
The American Society for Information Science (ASIS) is a nonprofit professional
association organized for scientific, literary, and educational purposes and dedicated to the creation, organization, dissemination, and application of knowledge concerning information and its transfer. Founded in the mid-1930s, ASIS
has a history which stems from the earliest days of the modern era of documentation. ASIS counts among its membership some 4,000 information specialists
from such fields as computer science, management, engineering, librarianship,
chemistry, linguistics, and education. ASIS and its members are called upon
to help determine new directions and standards for the development of information policies and practices. The mission of the American Society for
Information Science is to advance information professionals and the field of information science. [This ASIS is not to be confused with the American Society
for Industrial Security, a trade assoc. that works on commercial security issues,
largely to the detriment of worker privacy.]
General: asis@cni.org
Voice: +1 301 495 0900
Fax: +1 301 495 0810
Snail: P.O. Box 554, Washington, DC 20044-0554 USA

*Americans Communicating Electronically (ACE)*
ACE membership is diverse and represents private and government organizations and individuals who wish to promote interactive communications among
federal, state, and local governments, private businesses, public libraries and
schools, rural cooperatives, public and private universities, community-based
arts and theater groups, voluntary associations, job training services, and health
care organizations. The members of ACE are particularly concerned that access
and participation be made possible and convenient for Americans who do not
own modem-equipped computers. To support the development of interactive
communications between governments and communities, ACE is recommend-
ing that all government agencies establish information access programs to help

create and foster an "interactive citizen-government communications system." Many government agencies, from the White House to the NSF and the Dept. of Labor, are already participating in the ACE project. Unlike everything else on this list, ACE is actually a government-sponsored project. There are several ACE mailing lists: ACE-MG (general ACE info), CET-MG (Communities in Economic Transition), CET-NEWS (CET bulletins), etc.
Basic info: info@ace.esusda.gov
General: letters@ace.esusda.gov
Mailing list subscriptions: almanac@ace.esusda.gov, message body: "subscribe
   <listname> <Firstname> <Lastname>"

*Center for Civic Networking (CCN/CivicNet)*
The Center for Civic Networking is a nonprofit organization, based in Boston and Washington, D.C., that promotes broad public benefits of the emerging national information infrastructure. The Center brings together expertise in large-scale computer and network systems, community-based applications of computing, nonprofit management, community development, architecture, public policy, and democratic participation. The Center's Programs focus on framing a national vision for civic networking, developing a policy framework that supports civic networking, developing and supporting model civic networking projects, and assisting in the technology transfer needed to achieve the broad-based benefits of civic networking. CCN is involved with SDIN network, the Cambridge Civic Forum, the "From Townhalls to Local Civic Networks" conference, and ACE.
General: Miles Fidelman <mfidelman@world.std.com>
<mfidelman@civicnet.org>, Exec. Dir.
Richard Civille <rciville@civicnet.org>, Dir., D.C. office
FTP: world.std.com, ftp/amo/civicnet, ftp.eff.org, pub/Groups/CCN
Gopher: gopher.std.com, 1/associations/civicnet, gopher.eff.org, 1/Groups/CCN
www: gopher://gopher.std.com:70/11/associations/civicnet
http://www.eff.org/pub/Groups/CCN/
Voice: +1 202 362 3831 (R. Civille, Washington, D.C., office)

*Center for Democracy and Technology (CDT)*
CDT is a nonprofit, public interest organization. The mission of CDT is to develop public policy solutions that advance constitutional civil liberties and democratic values in new computer and communications media. The Center will pursue its mission through policy research, public education, and coalition

building. The Center will marshal legal, technical, and public policy expertise on behalf of civil liberties goals, including: maximizing free speech and the free flow of information online, giving citizens more control over personal information, protecting privacy online, and guaranteeing public access to electronic government information.

Info: info@cdt.org
General: ask@cdt.org
FTP: ftp.cdt.org, /pub/cdt/
Gopher: gopher.eff.org, 1/Groups/CDT (not full archive; see FTP/www sites.)
www: http://www.cdt.org/
Voice: +1 202 637 9800
Fax: +1 202 637 0968
Snail: 1001 G St. NW, Ste. 700E, Washington, DC 20001 USA

*Center for Governmental Studies (CGS)*
A Los Angeles–based nonprofit, research organization which works to improve the processes of media and democratic governance. In 1989 the Center launched the California Channel, the nation's first "state C-SPAN" now available to over 4 million California homes. With the Babcock, Carnegie, Cummings, Gerbode, and Irvine foundations, CGS initiated the Democracy Network, a plan for an interactive network/station for the NII that will allow voters, through their TV sets or computers, to access information on demand from political candidates and ballot measure committees, and to talk with candidates and voters in an electronic town hall, to increase voter participation, decrease campaign costs, encourage candidates to devote more attention to substantive issues, and demonstrate to elected officials the value of incorporating free voter information into the new definition of "universal service."

General: dnetcgs@aol.com
Voice: +1 310 470 6590
Fax: +1 310 475 3752
Snail: 10951 West Pico Blvd., Suite 206, Los Angeles, CA 90064 USA

*Computer Professionals for Social Responsibility (CPSR)*
CPSR is a national membership organization, based in Palo Alto, California. CPSR conducts many activities to protect privacy and civil liberties. Membership is open to the public and support is welcome. CPSR maintains 24 local chapters in the U.S., and has several international affiliates. CPSR hosts several mailing lists, including cpsr-cpu (CPSR's "CPU" newsletter for information

technology workers), cpsr-announce (CPSR's general news and announcements list, gated to Usenet newsgroup comp.org.cpsr.announce), and bawit-announce (Bay Area Women & Information Technology working group announcements), among others. CPSR sponsors an annual conference, maintains a large Internet archive site of information, and sponsors working groups on civil liberties and other issues.

General (nat'l. HQ): cpsr@csli.stanford.edu

Washington, D.C., chapter: Larry Hunter <hunter@nlm.nih.gov>

N.Y. chapter: David Friedlander <friedd@pipeline.com>

Berkeley, Calif., chapter: Karen Coyle <cpsr-berkeley@cpsr.org>

Palo Alto, Calif., chapter: Andre Bacard <abacard@well.sf.ca.us>

Portland, Oreg., chapter: Erik Nilsson <erikn@goldfish.mitron.tek.com>

Los Angeles chapter: Rodney J. Hoffman <rodney@oxy.edu>

Mailing lists: listserv@cpsr.org, message body: "subscribe <listname—see above> <firstname> <lastname>"

FTP: ftp.cpsr.org, /cpsr

Gopher: gopher.cpsr.org, 1/cpsr

www: http://www.cpsr.org/home

Usenet: comp.org.cpsr.talk, comp.org.cpsr.announce

Nanotechnology SIG: Ted Kaehler <kaehler2@applelink.apple.com>

Electoral issues: Eva Waskell, voice: +1 703 435 1283 evenings

Snail: CPSR National Office, P.O. Box 717, Palo Alto, CA 94302 USA

Voice: +1 415 322 3778

Fax: +1 415 322 3798

CPSR/Berkeley, SE P.O. Box 40361, Berkeley, CA 94704 USA

Voice: +1 415 398 2818

*Consortium for School Networking (CoSN)*
Through computer networking, the Consortium will help educators and students access information and communications resources that will increase their productivity, professional competence, and opportunities for learning and collaborative work. The Consortium advocates the following goals: the timely deployment of the national research and education network; the development and distribution of network-based information resources for schools; the development of the human resources needed to make full and efficient use of networks through staff development programs, educational materials and software. [The Consortium attempts to] form a national leadership group for educational tele-

communications, to have a voice in shaping policy in this area; provide access to information about the National Research and Education Network (NREN) and other educational telecommunications efforts; reach a large community of individuals involved in every aspect of network technology and its application to K-12 education; help advance the development of information resources and tools for networking; foster collaborative opportunities to develop new resources and services for educators.

CoSN is a nonprofit organization, 501(c)(3) application pending.
General: cosn@bitnic.bitnet, cosn%bitnic@cunyvm.cuny.edu
Gopher: digital.cosn.org
Snail: P.O. Box 6519, Washington, DC 20035-5193 USA
Voice: +1 202 466 6296
Fax: +1 202 872 4318

*Corporation for National Research Initiatives (CNRI)*
A nonprofit research and development organization formed in 1986 to help focus U.S. strengths in information processing technology. Working with industry, government, and academia, CNRI is engaged in scientific research on the design of an experimental infrastructure which can improve the country's long-range scientific and engineering productivity. CNRI organizes multi-party collaborative research activities among U.S. government, business, and academic organizations. An experimental information infrastructure will provide an important basis for joint university/industry research and facilitate rapid transfer of advanced scientific concepts and technology between research groups and also into experimental applications.
General: info@cnri.reston.va.us
Gopher: ietf.cnri.reston.va.us, 1/CNRI Information
www: gopher://ietf.cnri.reston.va.us:70/11/CNRI%20Information
Snail: 895 Preston White Drive, Suite 100, Reston, VA 22091 USA
Voice: +1 703 620 8990

*Corporation for Research and Educational Networking (CREN)*
Despite the name, this is a nonprofit organization. CREN advances the goals of institutions of higher education by facilitating, catalyzing, and leveraging contributions from the worldwide higher education community directed toward building a global computing and communications infrastructure that: supports access to shared information services and resources; supports scholarly collaboration and educational outreach; and contributes to enhanced individual and

institutional productivity. CREN provides BITNET (and thus Internet email) access to member institutions, and is also working on NII issues, hoping to help ensure that such a future network provides for the needs of the educational and research communities.

General: bitnet@cren.net
FTP: info.cren.net, cren.org
Gopher: info.cren.net
Snail: 1112 16th St. NW, Suite 600, Washington, DC 20036 USA
Voice: +1 202 872 4200

*Electronic Frontier Foundation (EFF)*

A nonprofit public interest membership organization, working to protect individual rights in the emerging information age. EFF supports legal and legislative action to protect the civil liberties of online users; hosts and participates in related conferences and projects, including Big Dummy's Guide to the Internet, and Computers and Academic Freedom; and works to educate the online community about its legal rights and responsibilities. EFF members receive online bulletins about the critical issues and debates affecting computer-mediated communications and participate in online political activism. Donations are welcome and are tax deductible. EFF is a 501(c)(3) tax-exempt organization.

Basic info: info@eff.org
General: ask@eff.org
Membership: membership@eff.org
Legal: Shari Steele <ssteele@eff.org>, Director of Legal Services; Mike Godwin <mnemonic@eff.org>, Staff Counsel
Policy/Open Platform/NII: Drew Taubman <drewt@eff.org>, Executive Director
Tech: Dan Brown <brown@eff.org>, Systems Administrator
Online newsletter: Stanton McCandlish <mech@eff.org>, Online Activist
Hardcopy publications: pubs@eff.org
Mailing list requests: listserv@eff.org, message body: "HELP" or "LONGINDEX"
FTP: ftp.eff.org
Gopher: gopher.eff.org
WAIS: wais.eff.org [temporarily unavailable for overhauling]
www: http://www.eff.org/
Usenet: comp.org.eff.talk, comp.org.eff.news, alt.politics.datahighway
WELL: g eff

AOL: keyword EFF
CIS: go effsig
Computers & Academic Freedom Project: kadie@eff.org, greeny@eff.org
CAF mailing list: listserv@eff.org (add comp-academic-freedom-news)
Computer Underground Digest Archives: cudarch@eff.org
Snail: 1667 K St. NW, Suite 801, Washington, DC 20006-1605 USA
Voice: +1 202 861 7700
Fax: +1 202 861 1258
BBS: +1 202 861 1223 (16.8k ZyXEL) +1 202 861 1224 (14.4k V.32bis)

*Electronic Privacy Information Center (EPIC)*
The Electronic Privacy Information Center is a public interest research center in Washington, D.C. It was established in 1994 to focus public attention
on emerging privacy issues relating to the National Information Infrastructure, such as the Clipper Chip, the Digital Telephony proposal, medical record
privacy, and the sale of consumer data. EPIC is sponsored by the Fund for Constitutional Government and Computer Professionals for Social Responsibility.
EPIC publishes the EPIC Alert and EPIC Reports, pursues Freedom of Information Act litigation, and conducts policy research on emerging privacy issues.
EPIC also works closely with Privacy International, a human rights group, on
domestic and international privacy issues.
General: info@epic.org
FTP: ftp.cpsr.org, /cpsr/privacy/epic/ ftp.cpsr.org, /cpsr/alert/
Gopher: gopher.cpsr.org, 1/cpsr/privacy/epic/ gopher.cpsr.org, 1/cpsr/alert
WWW: http://epic.digicash.com/epic
http://cpsr.org/dox/privacy.html
http://cpsr.org/cpsr/privacy/epic/
http://cpsr.org/cpsr/alert/
Snail: 666 Pennsylvania Ave. SE, Suite 301, Washington, DC 20003 USA
Voice: +1 202 544 9240
Fax: +1 202 547 5482

*Higher Education Information Resources Alliance (HEIRAlliance) — CNI*
The Higher Education Information Resources Alliance (HEIRAlliance) is a vehicle for cooperative projects between the Association of Research Libraries,
CAUSE, and EDUCOM. Currently, its major projects are the Coalition for Networked Information, (formed in 1990; promotes the creation of and access to
information resources in networked environments in order to enrich scholar-

ship and to enhance intellectual productivity. Roughly 175 organizations and institutions are members of the Coalition) and the HEIRAlliance Executive Strategies reports (designed to keep chief higher education executives informed about critical issues related to information technologies).

General: Craig A. Summerhill <craig@cni.org>, Systems Coord./Program Ofcr.

CNI general: Joan Lippincott <joan@cni.org>, Asst. Exec. Dir.

Exec. Strategies report queries: Karen McBride <kmcbride@CAUSE.colorado.edu>

CNI Announcements list: listproc@cni.org, message body: "subscribe cni-announce <firstname> <lastname>"

Gopher: gopher.cni.org

www: gopher://gopher.cni.org:70/1

Snail: 21 Dupont Circle NW, Washington, DC 20036 USA

Voice: +1 202 296 5098

Fax: +1 202 872 0884

*League for Programming Freedom (LPF)*

The League is a membership-based organization whose aim is to bring back the freedom to write software. The League is not opposed to the legal system that Congress intended—copyright on individual programs. Our aim is to reverse the recent changes made by judges in response to special interests, often explicitly rejecting the public interest principles of the Constitution. The primary areas of League activity are software patents and user interface copyright.

General: lpf@uunet.uu.net

FTP: prep.ai.mit.edu, /pub/lpf/

www: http://www.lpf.org/

Voice: +1 617 621 7084

Snail: 1 Kendall Square #143, P.O. Box 9171, Cambridge, MA 02139 USA

*National Public Telecomputing Network (NPTN)—a.k.a. Free-Net*

The National Public Telecomputing Network exists to make free public access to computerized communications and information services a reality; to help people in cities throughout the U.S. and the world to establish free, open access, community computer systems (Free-Nets); to link those systems together into a common network similar to National Public Radio or PBS on TV; to help supplement what the local systems are able to produce with high quality network-wide services and features. NPTN is a 501(c)(3) nonprofit corporation. [Note: "Free-Net" is a service mark of the National Public Telecomputing

Network (NPTN), registered in the U.S. and other countries. While not all community networks are a part of the NPTN family, NPTN and its Affiliates and Organizing Committees represent the first and only international network of community computing systems.]

General: info@nptn.org [note: this is not an automailer, but a person]

Canadian Free-Net mailing list: listprocessor@cunews.carleton.ca (subscribe
  CAN-FREENET <name>)

FTP: nptn.org, /pub/nptn/

www: http://www.nptn.org/

Snail: National Public Telecomputing Network, P.O. Box 1987, Cleveland, OH
  44106 USA

Voice: +1 216 498 4050

Fax: +1 216 498 4051

NPTN Free-Net Affiliates Community Computer Systems modem numbers

| | | |
|---|---|---|
| Big Sky Telegraph | Dillon, Montana | +1 406 683 7680 |
| Buffalo Free-Net | Buffalo, New York | +1 716 645 6128 |
| Cleveland Free-Net | Cleveland, Ohio | +1 216 368 3888 |
| COIN | Columbia, Missouri | +1 314 884 7000 |
| Denver Free-Net | Denver, Colorado | +1 303 270 4865 |
| Lorain County Free-Net | Elyria, Ohio | +1 216 366 9721 |
| Medina County Free-Net | Medina, Ohio | +1 216 723 6732 |
| National Capital Free-Net | Ottawa, Ont., Canada | +1 613 564 3600 |
| Tallahassee Free-Net | Tallahassee, Florida | +1 904 576 6330 |
| Tristate Online | Cincinnati, Ohio | +1 513 579 1990 |
| Tri-Cities Free-Net | Hanford, Washington | +1 509 375 1111 |
| Victoria Free-Net | Victoria, B.C., Canada | +1 604 595 2300 |
| Wellington Citynet | Wellington, New Zealand | +64 4 801 3060 |
| Youngstown Free-Net | Youngstown, Ohio | +1 216 742 3072 |

For more detailed information, including Internet addresses and login instructions, see ftp.eff.org, pub/Groups/NPTN-Freenet/login.info [Note: There are many other community network access providers, both NPTN and independent, and more are created every month.]

## OMB *Watch*

OMB Watch is a nonprofit research, educational, and advocacy organization that monitors Executive Branch activities affecting nonprofit, public interest, and community groups. OMB (the White House Office of Management and

Budget) is the main focus as it oversees nearly all executive branch functions. Our goal is to encourage broad public participation in government decision-making to promote a more open and accountable government. Our activities include: technical assistance on budget, regulatory accountability, government secrecy, and general government decision-making through publications, training sessions, and direct links to certain government data; community forums on the federal budget to reorder priorities to domestic needs; RTK NET (Right-to-Know computer network); advocacy—through the tools to empower community groups and coordination of coalitional efforts in a variety of areas; support of public access to and use of government information. Most activity is conducted offline, so send a snailmail address if you want OMB Watch materials.
General: Patrice McDermot <patricem@cap.gwu.edu>

*Telecommunications Policy Roundtable (TPR)*
The TPR is a coalition of more than 100 organizations which was organized in the spring of 1993 to discuss federal telecommunications and information policy. The group as a whole meets every month in Washington, D.C., and it also sponsors several regular and ad hoc committees and subcommittees to address specific roundtable concerns. TPR members include EFF, CME, CNI, TAP, and CPSR. TPR sponsors an open-to-all mailing list forum for discussion of U.S. telecom policy, called ROUNDTABLE.
General: Jeff Chester <cme@access.digex.net> Coralee Whitcomb <cwhicom@bentley.edu> Paul Johnson <pjbrady@delphi.com>
Technical admin: Craig Summerhill <craig@cni.org>
Mailing list subscription: listproc@cni.org, message body: "SUBSCRIBE ROUND-
    TABLE <firstname> <lastname>"
Mailing list admin: Jamie Love <love@essential.org>
Voice: +1 202 628 2620 (Center for Media Education, initial contact for organi-
    zations wishing to join TPR)
Fax: +1 202 234 5176 (c/o Jamie Love at TAP)

*US Privacy Council (USPC)*
A coalition of U.S. privacy groups and individuals founded in 1991 to deal with privacy issues in the U.S. USPC works in Washington, D.C., monitoring legislation and the activities of government agencies. USPC works closely with other groups on privacy issues including National ID cards, reforming credit reporting, Caller ID, and international issues.
General: privtime@access.digex.net

Snail: P.O. Box 15060, Washington, DC 20003 USA
Voice: +1 202 829 3660

*Voters Telecomm Watch (VTW)*

The Voters Telecomm Watch is a volunteer organization dedicated to monitoring federal legislation that affects telecommunications and civil liberties. VTW is based primarily out of New York, though they have volunteers throughout the U.S. Voters Telecomm Watch keeps scorecards on legislators' positions on legislation that affects telecommunications and civil liberties.

General: vtw@vtw.org
Admin: shabbir@panix.com
Press contact: stc@vtw.org
Mailing list requests (announcements): listproc@vtw.org, message body: "subscribe vtw-announce <firstname> <lastname>"
Mailing list requests (discussion): listproc@vtw.org, message body: "subscribe vtw-list <firstname> <lastname>"
FTP: ftp.eff.org, /pub/Groups/VTW/ [minimal info]
Gopher: gopher.panix.com, 1/vtw [much more info than at ftp site]
www: gopher://gopher.panix.com/11/vtw
Voice: +1 718 596 2851

## Administrivia

*Who/Where*

This list is maintained by Stanton McCandlish <mech@eff.org>, and faq-ized and distributed by L. Detweiler. It is based on a previous version by Shari Steele. Future updates will be posted to ACTION mailing list and comp.org.eff.talk, besides several other places. The most current version is available from:
ftp.eff.org, /pub/EFF/Issues/Activism/activ_groups.faq
gopher.eff.org, 1/EFF/Issues/Activism, activ_groups.faq
http://www.eff.org/pub/EFF/Issues/Activism/activ_groups.faq

*Scope*

This list focuses on:

1. organizations dealing with online issues such as cryptography, intellectual freedom in networking, and access to government information,
2. organizations working on access to online resources, and supporting community networking, and
3. general activism and civil liberties organizations providing material online.

The listing of all activism groups (e.g., [non-]smokers' rights, environmentalism, gun ownership/regulation, etc.) and partisan politics groups is outside the scope of this list.

### Distribution

This FAQ is Copyright 1995 Electronic Frontier Foundation, and is made available as a freeware service to the online community, on behalf of the ACTION forum.

### The Electronic Frontier Foundation

The Electronic Frontier Foundation (EFF) was founded in July of 1990 to ensure that the principles of privacy, open access, and freedom of expression are protected as new communications technologies emerge. Since its inception, EFF has worked to shape our nation's communications infrastructure and the policies that govern it in order to maintain and enhance intellectual freedom, privacy, and other democratic values. We believe that our overriding public goal must be the creation of Electronic Democracy.

Membership & General Info: info@eff.org

The Electronic Frontier Foundation, 1667 K St. NW, Suite 801, Washington, DC 20006-1605 USA

+1 202 861 7700 (voice) +1 202 861 1258 (fax) +1 202 861 1223 (BBS—16.8k ZyXEL) +1 202 861 1224 (BBS—14.4k V.32bis)

Internet: ask@eff.org

### Action: The Activism Online Forum

Action is an Internet "mailing list" forum, and serves as a virtual community supporting grassroots political action through networking technology. Action is a focused working group, rather than a chat area. To subscribe, send a message body of "subscribe ACTION" to listserv@eff.org via Internet email.

# Notes

### The Immersions: A Preface

The following sources were consulted:

Brook, James, and Iain A. Boal, eds., *Resisting the Virtual Life: The Culture and Politics of Information* (San Francisco: City Lights Books, 1995), ix.

Chapman, Gary, "Making Sense Out of Nonsense: Rescuing Reality from Virtual Reality," in *Culture on the Brink: Ideologies of Technology*, ed. Gretchen Bender and Timothy Druckery (Seattle: Bay Press, 1994), 149–155.

Dyson, Esther, George Gilder, George Keyworth, and Alvin Toffler, "Cyberspace and the American Dream: A Magna Carta for the Knowledge Age," release 1.2 (August 22, 1994), available via e-mail info@bionomics. org.

Katz, Jon, "Return of the Luddite," *Wired* 3.06 (June 1995): 165.

Keegan, Paul, "The Digerati!" *New York Times Magazine* (May 21, 1995): 38.

Kleinfield, N. R., "Stepping Through a Computer Screen, Disabled Veterans Savor Freedom," *New York Times* (March 12, 1995): 39.

Talbott, Stephen, *The Future Does Not Compute: Transcending the Machines in Our Midsts* (Sebastopol, Calif.: O'Reilly, 1995).

Winner, Langdon, *The Whale and the Reactor* (Chicago: University of Chicago Press, 1986).

## Chapter 1: Real Cold, Simulated Heat

1. Baudrillard, Jean, *America,* trans. Chris Turner (New York: Verso, 1988), 37: "The ecstasy of the polaroid. . . : to hold the object and its image almost simultaneously as if the conception of light of ancient physics or metaphysics, in which each object was thought to secrete doubles or negatives of itself that we pick up with our eyes, has become a reality. It is a dream. It is the optical materialization of a magical process. The polaroid photo is a sort of ecstatic membrane that has come away from the real object."

2. Bellah, Robert N., Richard Madsen, William M. Sullivan, Ann Swidler, and Steven M. Tipton, *Habits of the Heart: Individualism and Commitment in American Life* (New York: Harper and Row, 1985).

3. Tocqueville, Alexis de, *Democracy in America,* ed. J. P. Mayer, trans. George Lawrence (New York: Harper Perennial, 1988), 508. Italics added.

4. Markoff, John, "The Lost Art of Getting Lost," *New York Times* (September 18, 1994): sec. 4, 1.

5. Baudrillard, 32: "Everything is destined to reappear as simulation. Landscapes as photography, women as the sexual scenario, thoughts as writing, terrorism as fashion and the media, events as television. Things seem only to exist by virtue of this strange destiny. You wonder whether the world itself isn't just here to serve as advertising copy in some other world."

6. Pruitt, Steve, and Tom Barrett, "Corporate Virtual Workspace," in *Cyberspace: First Steps,* ed. Michael Benedikt (Cambridge: MIT Press, 1991), 383–409.

7. Pruitt and Barrett, 391–392.

8. Maney, Kevin, "High-Tech Rooms with 3-D View," *USA Today* (May 11, 1995): sec. B, 1, 2.

9. Barron, James, "A New Species of Couch Potato Takes Root," *New York Times* (November 6, 1994): sec. 2, 1.

10. Kroker, Arthur, and Michael A. Weinstein, *Data Trash* (New York: St. Martin's Press, 1994): 7.

11. Heim, Michael, *The Metaphysics of Virtual Reality* (New York: Oxford University Press, 1993), 136.

12. Kroker and Weinstein, 7: "The virtual class has seized the imagination of contemporary culture by conceiving a techno-utopian high-speed cybernetic grid for travelling across the electronic frontier. In this mythology of the new technological frontier, contemporary society is either equipped for

fast travel down the main arterial lanes of the information highway, or it simply ceases to exist as a functioning member of technotopia."

13. Markoff, 1.

14. Livingston, John A., *Rogue Primate: An Exploration of Human Domestication* (Toronto: Key Porter Books, 1994), 14, 57, 13.

## Chapter 2: Immersive Virtualists and Wired Communitarians

1. Mitchell, William J., *City of Bits: Space, Place, and the Infobahn* (Cambridge: MIT Press, 1995), 29–30. For another extensive discussion of our cyborgian destiny see Stone, Allucquere Rosanne, *The War of Desire and Technology at the Close of the Mechanical Age* (Cambridge: MIT Press, 1995).

2. Gruber, Michael, "Neurobotics," *Wired* 2.10 (October 1994): 111; on Brooks and his team see Freedman, David H., "Bringing Up RoboBaby," *Wired* 2.12 (December 1994): 74.

3. Rheingold, Howard, *The Virtual Community: Homesteading on the Electronic Frontier* (Reading, Mass.: Addison-Wesley, 1993), 7.

4. Barlow, John Perry, "The Economy of Ideas," *Wired* 2.03 (March 1994): 84; Doheny-Farina, Stephen, letter to "Rants and Raves," *Wired* 2.06 (June 1994): 22.

5. I make this argument in detail in Doheny-Farina, Stephen, *Rhetoric, Innovation, Technology: Case Studies of Technical Communication in Technology Transfers* (Cambridge: MIT Press, 1992).

6. Barlow, John Perry, response to Stephen Doheny-Farina, "Rants and Raves," *Wired* 2.06 (June 1994): 22.

7. Rushkoff, Douglas, *Media Virus* (New York: Ballantine Books, 1994), 31.

8. Rushkoff, 21.

9. Rushkoff, 218. Rushkoff points out that critics like Kroker were born too early to understand the medianet. "Kroker's brilliant but misguided analysis is typical of his generation of philosophers who, growing up before the advent of mass media, have only the tools to observe media but not the language or translation skills to partake in it. . . . the inferences he draws totally ignore the nature of the new and growing relationship between our lives and our media. As long as we view media, or technology for that matter, as something separate from ourselves—something unnatural—we will always see it as the enemy to the natural unfolding of our culture."

10. Rheingold, 23.

11. Baudrillard, Jean, "Simulacra and Simulations," in *Selected Writings*, ed.

Mark Poster (Stanford: Stanford University Press, 1988), 166–184. Virtual community is like Disneyland, and placed communities, like America: "Disneyland is there to conceal the fact that it is the 'real' country, all of 'real' America, which *is* Disneyland. . . . Disneyland is presented as imaginary in order to make us believe that the rest is real, when in fact all of Los Angeles and the America surrounding it are no longer real, but of the order of the hyperreal and of simulation. It is no longer a question of a false representation of reality . . . but of concealing the fact that the real is no longer real" (172).

12. Berry, Wendell, *Sex, Economy, Freedom, and Community* (New York: Pantheon Books, 1993), 14–15.

13. For example, the 150th anniversary issue of the *Economist* (January 1994).

14. Drucker, Peter, "The Age of Social Transformation," *Atlantic Monthly* 274, no. 5 (November 1994): 53–80. All quotations in the following discussion of Drucker's views are from this article.

15. Kemmis, Daniel, *Community and the Politics of Place* (Norman: University of Oklahoma Press, 1990), 138. All quotations in the following discussion of Kemmis's views are from this book.

16. Sanders, Scott Russell, *Staying Put: Making a Home in a Restless World* (Boston: Beacon Press, 1993), 117.

17. After presenting this image, Baudrillard says that it reminds him of some American university campuses. Baudrillard, Jean, *America*, trans. Chris Turner (New York: Verso, 1988), 44.

18. McKibben, Bill, *The Age of Missing Information* (New York: Plume, 1992).

19. Borgmann, Albert, *Crossing the Postmodern Divide* (Chicago: University of Chicago Press, 1992), 126.

**Chapter 3: Virtual Vermont**

1. Schwartz, Evan, "Linking the Information Superhighway to Main Street," *New York Times* (October 9, 1994): sec. 3, 10.

2. Bellah, Robert N., Richard Madsen, William M. Sullivan, Ann Swidler, and Steven M. Tipton, *Habits of the Heart: Individualism and Commitment in American Life* (New York: Harper and Row, 1985), 72.

3. Bellah et al., 72.

4. Kemmis, Daniel, *Community and the Politics of Place* (Norman: University of Oklahoma Press, 1990), 4–8.

5. Oldenburg, Ray, *The Great Good Place* (New York: Paragon House, 1989).

6. For a discussion of the distinctions between public and private life, distinctions both real and false, see Mouffe, Chantal, "Democratic Citizenship and the Political Community," in *Community at Loose Ends,* ed. Miami Theory Collective (Minneapolis: University of Minnesota Press, 1991), 70–82.

## Chapter 4: Seeking Public Space in a Virtual World

1. Bruckman, Amy, and Mitchel Resnick, "Virtual Professional Community: Results from the MediaMOO Project" (presented at the Third International Conference on Cyberspace, Austin, Tex., May 15, 1993), available via ftp media.mit.edu /pub/asb/papers.
2. Rheingold, Howard, "The Future of MUDs," *Wired* 1.3 (July–August 1993): 72; Bennahum, David, "Fly Me to the MOO: Adventures in Textual Reality," *Lingua Franca* (June 1994): 29.
3. Bruckman and Resnick, 5; Negroponte, Nicholas, *Being Digital* (New York: Alfred A. Knopf, 1995), 182.
4. Pruitt, Steve, and Tom Barrett, "Corporate Virtual Workspace," in *Cyberspace: First Steps,* ed. Michael Benedikt (Cambridge: MIT Press, 1991), 383–409.
5. For these and other features see, for example, no. 6927, Vlad's Hug Feature, and no. 9610, Gracie's Big Ass Spam Features, at MediaMOO.
6. Unsworth, John, "Living Inside the (Operating) System: Community in Virtual Reality," in *Computer Networking and Scholarly Communication in the Twenty-First Century,* ed. Teresa M. Harrison and Timothy D. Stephen (Albany: SUNY Press, in press); see also Selfe, Cynthia, and Richard Selfe, "The Politics of the Interface," *College Composition and Communication* 45, no. 4 (December 1994): 480–504.
7. See, for example, Reid, Elizabeth, "Electropolis: Communication and Community on Internet Relay Chat," B.A. honors thesis, University of Melbourne, Australia, 1991; Reid, Elizabeth, "Cultural Formations in Text-Based Virtual Reality," master's thesis, University of Melbourne, Australia, 1994, available via ftp.eff.org /pub/cud/papers. See also Reid, Elizabeth, "Virtual Worlds: Culture and Imagination," in *Cybersociety: Computer-Mediated Communication and Community,* ed. Steven G. Jones (Thousand Oaks, Calif.: Sage, 1995), 164–183.
8. Lea, Martin, and Russell Spears, "Love at First Byte? Building Personal Relationships over Computer Networks," in *Understudied Relationships: Off*

*the Beaten Track,* ed. Julia T. Wood and Steve Duck (Beverly Hills, Calif.: Sage, in press). See also Spears, Russell, and Martin Lea, "Panacea or Panopticon? The Hidden Power in Computer-Mediated Communication," *Communication Research* 21, no. 4 (August 1994): 427–459.

9. For an examination of racial bias online see Selfe and Selfe.

10. Dickel, M. H., "Bent Gender: Virtual Disruptions of Gender and Sexual Identity," *Electronic Journal of Communication* special issue, Networked Virtual Realities and Communication, ed. S. Doheny-Farina (September 1995).

11. Truong, Hoai-An (with additional writing and editing by Gail Williams, Judi Clark, and Anna Couey in conjunction with members of BAWIT—Bay Area Women in Telecommunications), "Gender Issues in Online Communication," version 4.2, available via http://english.hss.cmu.edu.

12. For discussions of gender-bending via the net see Bruckman, A. S., "Gender Swapping on the Internet" (presented at INET '93), available via ftp media.mit.edu /pub/asb/papers; Bruckman, A. S., Pavel Curtis, Cliff Figallo, and Brenda Laurel, "Approaches to Managing Deviant Behavior in Virtual Communities," Association for Computing Machinery, 1994, available via ftp media.mit.edu /pub/asb/papers; Curtis, Pavel, "Mudding: Social Phenomena in Text-Based Virtual Realities" (presented at DIAC '92), available via ftp parcftp.xerox.com /pub/MOO/papers.

13. Dibbel, Julian, "Rape in Cyberspace," *Village Voice* 38 (1993): 51, available via gopher eff.org; Quittner, Josh, "Johnny Manhattan Meets the FurryMuckers," *Wired* 2.03 (March 1994): 92. See also Stone, Allucquere Rosanne, *The War of Desire and Technology at the Close of the Mechanical Age* (Cambridge: MIT Press, 1995), 171.

14. A transcript of the Cyberspace and the Humanities Conference is available via ftp parcftp.xerox.com.

15. Netoric archives can be found via http://www.cs.bsu.edu/homepages/ siering/netoric.html.

16. Bruckman and Resnick, 6.

17. Pt MOOt papers and data are available via gopher actlab.rtf.utexas.edu /Virtual Environments/MOO/Point MOOt.

18. Oldenburg, Ray, *The Great Good Place* (New York: Paragon House, 1989), 210.

## Chapter 5: Seeking Public Space on the Internet

1. Winner, Langdon, "A Victory for Computer Populism," *Technology Review* (May–June 1991): 66.
2. Lewis, Peter H., "On the Internet, Dissidents' Shots Heard 'Round the World,'" *New York Times* (June 5, 1994): sec. 4, 18.
3. Kantrowitz, Barbara, "Dissent on the Hard Drive," *Newsweek* (June 27, 1994): 59.
4. Gonzalez, David, "The Computer Age Bids Religious World to Enter," *New York Times* (July 24, 1994); Lewis, Peter H., "Voters and Candidates Meet on Information Superhighway," *New York Times* (November 4, 1994): sec. 1, 30; Gleick, James, "The Information Future: Out of Control (and It's a Good Thing," *New York Times Magazine* (May 1, 1994): 57; Strangelove, Michael, "The Geography of Consciousness: Cyberspace and the Changing Landscape of the Self," *WAVE* 4 (October 1994): 37–38.
5. Gurak, Laura, "The Rhetorical Dynamics of a Community Protest in Cyberspace: The Case of Lotus MarketPlace," Ph.D. diss., Rensselaer Polytechnic Institute, 1994.
6. Gurak, 47.
7. Gurak, 99–100.
8. Barber, Benjamin, keynote address (presented at DIAC '94, Conference of the Computer Professionals for Social Responsibility, Cambridge, Mass., 1994).
9. Gurak, 80.
10. Savan, Leslie, *The Sponsored Life: Ads, TV, and American Culture* (Philadelphia: Temple University Press, 1994), 8–9.
11. Domzal, Teresa J., and Jerome B. Kernan, "Mirror, Mirror: Some Postmodern Reflections on Global Advertising," *Journal of Advertising* 22, no. 4 (December 1993): 1–20. All quotations of Domzal and Kernan are from this article.
12. Domzal and Kernan, 6, 8.
13. Domzal and Kernan, 8. They refer to Schiller, Herbert, *Culture, Inc.: The Corporate Takeover of Public Expression* (New York: Oxford University Press, 1989).
14. Maes, Pattie, "Agents that Reduce Work and Information Overload," *Communications of the ACM* 37, no. 7 (July 1994): 31. See also Berkun, Scott,

"Agent of Change," *Wired* 3.04 (April 1995): 117; and Rheinhardt, Andy, "The Network of Smarts," *Byte* 19, no. 10 (October 1994): 51–60.

15. King, Thomas, "Zap! Advertisers Need New Ways to Reach Consumers in the Interactive Age. Are They Up to the Challenge?" *Wall Street Journal* (September 9, 1994): R12; Savan, 49.

16. Lewis, Peter, "On-Line Ads to Tempt the Hungry Computer Nerd," *New York Times* (August 31, 1994): sec. O, 4.

17. Quotations are from Zachary, G. Pascal, "Advertisers Anticipate Interactive Media as Ingenious Means to Court Consumers," *Wall Street Journal* (August 17, 1994): B1.

18. Ehrenfeld, David, "Pseudocommunities," *Orion* (Autumn 1993): 5–6.

## Chapter 6: Telecommuting

1. Strand, Patricia, "Work at Home, Shop at Home," *Advertising Age* (January 10, 1994): S7.

2. Ravitch, Diane, "When School Comes to You," *Economist* (January 1994): 43.

3. Fisher, Lawrence M., "Quakes Four Years Apart Show How Far Telecommuting Has Come," *New York Times* (February 13, 1994): sec. 3, 10; Horton, Cleveland, "Calif. Shaken, but Barely Stirred," *Advertising Age* (July 25, 1994): 12.

4. Strand, S7. See also Greengard, Samuel, "Making the Virtual Office a Reality," *Personnel Journal* (September 1994): 71.

5. Greengard, 71.

6. On Compaq see Sullivan, R. Lee, "The Office That Never Closes," *Forbes* (May 23, 1994): 212–213. On Gemini see Greengard, 71. On Perkin-Elmer see Shellenbarger, Sue, "Overwork, Low Morale Vex the Mobile Office," *Wall Street Journal* (August 17, 1994): B1.

7. Sullivan, 213.

8. Kotkin, Joel, "Commuting via Information Superhighway," *Wall Street Journal* (January 27, 1994): A14. Subsequent references to Kotkin are from this article.

9. Davidow, William H., and Michael S. Malone, *The Virtual Corporation* (New York: Harper Business, 1992), 136–137.

10. Barnet, Richard J., and John Cavanaugh, *Global Dreams: Imperial Corporations and the New World Order* (New York: Simon and Schuster, 1994), 334, 337.

11. Shellenbarger, Sue, "I'm Still Here! Home Workers Worry They're Invisible," *Wall Street Journal* (December 16, 1993): B1.
12. Bowers, Brent, "Work-at-Home Deals Help Create New Entrepreneurs," *Wall Street Journal* (January 10, 1994): B2.
13. Greengard, 78.
14. Greengard, 78.
15. Shellenbarger, "Overwork," B1; Sullivan, 212–213.
16. Noble, Barbara Presley, "Electronic Liberation or Entrapment?" *New York Times* (June 15, 1994): sec. D, 4.
17. Greengard, 78.

## Chapter 7: Default Equals Offline

1. McCarthy, Patrick, "Traditional Labor Laws Apply to the Non-traditional Office," *Personnel Journal* (September 1994): 75.
2. DIAC '94, Conference of the Computer Professionals for Social Responsibility, Cambridge, Mass., 1994.
3. Rheingold, Howard, "PARC Is Back," *Wired* 2.02 (February 1994): 90–95. All quotations in the following discussion of Ubicomp are from this article.
4. Marx, Gary, "Privacy and Technology," *The World and I* (September 1990), quoted in Rheingold, Howard, *The Virtual Community: Homesteading on the Electronic Frontier* (Reading, Mass.: Addison-Wesley, 1993), 294.
5. Doheny-Farina, Stephen, "Default = Offline or Why Ubicomp Scares Me," *Computer-Mediated Communication Magazine* 1, no. 6 (October 1994), available via http://sunsite.unc.edu/cmc/mag/october.
6. Weiser, Mark, "The Technologist's Responsibilities and Social Change," *Computer-Mediated Communication Magazine* 2, no. 4 (April 1, 1995), available via http://sunsite.unc.edu/cmc/mag/april. Subsequent quotations of Weiser are from this article.
7. Porush, David, "Ubiquitous Computing vs. Radical Privacy: A Reconsideration of the Future," *Computer-Mediated Communication Magazine* 2, no. 3 (March 1, 1995), available via http://sunsite.unc.edu/cmc/mag/march. Subsequent quotations of Porush are from this article.
8. Wayne, Leslie, "If It's Tuesday, This Must Be My Family," *New York Times* (May 14, 1995): sec. 3, 1.

## Chapter 8: Virtual Schools

1. Diversity University MOO can be reached via telnet to moo.du.org 8888. Log on as guest.
2. Anderson, Christopher, "Cyberspace Offers Chance to Do 'Virtually' Real Science," *Science* 264 (May 13, 1994): 900–901.
3. Tiffin, John, "The Electronic Super Highway: International Perspectives" (panel discussion, International Council for Distance Education, Standing Conference of Presidents [ICDE SCOP], Second Annual Meeting, Saratoga Springs, N.Y., October 1994). Subsequent references to Tiffin are from this paper. See also Tiffin, John, and Lalita Rajasingham, *In Search of the Virtual Class: Education in an Informal Society* (London: Routledge, 1994).
4. "Policy Paper on Open and Distance Learning: Draft Summary" (International Council for Distance Education Paper for UNESCO, 10/14/94 draft version, presented to the International Council for Distance Education, Standing Conference of Presidents [ICDE SCOP], Second Annual Meeting, Saratoga Springs, N.Y., October 1994).
5. Kato, Hidetoshi, *National Institute of Multimedia Education Catalogue* (National Institute of Multimedia Education, Chiba-Shi, Japan, fax 043-275-5117).
6. Stemer, Rosalie, "The Virtual Classroom," *New York Times* (January 8, 1995), 4A, Education Life: 39.
7. Hetrick, Robert C., "Promoting Information Technology: A Seven-Step Approach," *AGB Priorities,* Association of Governing Boards 1 (Spring 1994): 11.
8. This is the argument I presented in Doheny-Farina, Stephen, *Rhetoric, Innovation, Technology: Case Studies of Technical Communication in Technology Transfers* (Cambridge: MIT Press, 1992). The argument is based in part on Dobrin, David, *Writing and Technique* (Urbana, Ill.: National Council of Teachers of English, 1989), 60–73.
9. Both of the following articles appeared in a special journal issue that I edited. The issue was devoted to networked virtual realities (MOOs, MUDs, MUSHs, and so on) and communication, and is available via e-mail comserve@vm.its.rpi.edu. Harris, Leslie D., "Dante in MOO Space: Using Networked Virtual Reality to Teach Literature," *Electronic Journal of Communication* 5, no. 4 (September 1995); Dykes, Merry, and Jennifer Waldorf, "Educational Telecommunication Usage in an After School Environment:

Using Recreational Practices Towards Educational Goals," *Electronic Journal of Communication* 5, no. 4 (September 1995).

10. Associated Press, "Living Textbook Project Debuts at Four N.Y. Schools," *Watertown (N.Y.) Daily Times* (May 6, 1995): 6.

11. Babu, S. V., Don H. Rasmussen, and Ian I. Suni, co-principal investigators, "Thin Film Technologies: Combined Research-Curriculum Development" (National Science Foundation ENG CRCD Program, 1994–96, funded at $360,000); Kane, James, Stephen Doheny-Farina, William Karis, and Jeffrey Layton, "Effective Engineering and Design Courseware via Boundary Methods" (National Science Foundation ERDTI Program, 1994–95, funded at $25,000).

12. Stemer, 39.

13. Bromley, Hank, "The Social Context of Educational Computing," *CPSR Newsletter* 12, no. 2 (Spring 1994): 7.

14. Apple, Michael W., "Computers and the Deskilling of Teaching," *CPSR Newsletter* 12, no. 2 (Spring 1994): 3.

## Chapter 9: The Communitarian Vision

1. Hinds, Michael deCourcy, "Among Amish, Suspect in Arson Is Well Known," *New York Times* (November 25, 1993): sec. A, 16. See also Associated Press, "Man Gets Ten Years for Amish Barn Fires," *New York Times* (June 9, 1994): sec. A, 16.

2. Hinds, sec. A, 16.

3. Terry, Don, "Gangs: Machiavelli's Descendants," *New York Times* (September 18, 1994): 26.

4. Clarkson University and Partners (principal authors: Ronald Chorba and Wm. Dennis Horn), "North Country Network Project: A Regional Partnership for Rural Access to Technology" (submitted to Telecommunications and Information Infrastructure Assistance Program [TIIAP], National Telecommunications and Information Administration, U.S. Department of Commerce, April 20, 1995).

5. Four brief but related histories of community networking are Beamish, Anne, "Communities On-Line: Community-Based Computer Networks," master's thesis, Massachusetts Institute of Technology, 1995, available via http://alberti.mit.edu/arch/4.207/anneb/thesis/toc.html; Morino, Mario, "Assessment and Evolution of Community Networking" (paper presented at the Apple Conference on Building Community Computing

Networks, Cupertino, Calif., May 5, 1994), available via http://www. morino.org; Schuler, Doug, "Community Networks: Building a New Participatory Medium," *Communications of the ACM* (January 1994): 39–51; Rheingold, Howard, *The Virtual Community: Homesteading on the Electronic Frontier* (Reading, Mass.: Addison-Wesley, 1993), 241–275.

6. Grundner, Tom, "Seizing the Infosphere: Toward the Formation of a Corporation for Public Cybercasting" (paper presented at DIAC '94, Conference of the Computer Professionals for Social Responsibility, Cambridge, Mass., 1994).

7. Morino.

8. Millennium Communications Group, "Communications as Engagement: A Communications Strategy for Revitalization," (Millennium Report to the Rockefeller Foundation, April 1994), available via gopher.civic.net/ Communication as Engagement.

9. Schuler, Doug, "Community Networks: Building a New Participatory Medium," *Communications of the ACM* (January 1994): 39.

10. Fidelman, Miles R., "Sustainable Development Information Network (SDIN): Report on Year 1," available via gopher info.hed.apple.com /Apple Library Users Group/Apple Library of Tomorrow/Community Networks/SDIN Year 1.

11. Civille, Richard, "Building Community Information Infrastructure: Universal Service for the Information Age," available via gopher info.hed.apple. com /Apple Library Users Group/Apple Library of Tomorrow/Community Networks/Building CII.

12. Beamish, chap. 1.

13. "Community Computing and the National Public Telecomputing Network" (National Public Telecomputing Network document), available via ftp nptn.org /pub/info.nptn/basic.guide.txt.Background.

14. "National Public Telecomputing Network Wins Federal Grant to Spur Development of Rural Community Networks" (National Public Telecomputing Network document, October 12, 1994), available via ftp nptn.org /pub/info.nptn.

15. "National Public Telecomputing Network Affiliate Systems and Organizing Committees" (National Public Telecomputing Network document, April 27, 1995), available via ftp nptn.org /pub/info.nptn.

16. National Capital Free-Net, "A Brief History of the National Capital Free-Net Project," available via telnet freenet.carleton.ca /main menu/About the

National Capital FreeNet/NCF Position Papers, By-Laws, History, Proposals/A Brief History of the National Capital FreeNet Project.

17. "National Public Telecomputing Network Wins Federal Grant."

18. Clarkson University and Partners.

19. Morino Institute, "The Promise and Challenge of a New Communications Age" (1995), available via http://www.morino.org.

20. Bishop, Ann P., and Scott Patterson, "A Pilot User Study of the Blacksburg Electronic Village," in *Navigating the Networks, Proceedings of the ASIS Mid-Year Meeting,* ed. Deborah Andersen, Thomas Garvin, and Mark Giguere (American Society for Information Science, 1994), 18–42.

21. Kavanaugh, Andrea, and Scott Patterson, "January 1994 Survey," available via http://www.bev.net/about bev page/registration, research, technical reports, publications/.

22. "A View of Blacksburg of the Future," available via http://www.bev.net.

23. "A View of Blacksburg of the Future."

24. Patrick, Andrew S., Alex Black, and Thomas E. Whalen, "Rich, Young, Male, Dissatisfied Computer Geeks? Demographics and Satisfaction from the National Capital FreeNet" (paper presented at Telecommunities '95: The International Community Networking Conference, Victoria, British Columbia, August 19–23, 1995).

## Chapter 10: Challenges to Community Networks

1. Beamish, Anne, "Communities On-Line: Community-Based Computer Networks," master's thesis, Massachusetts Institute of Technology, 1995, available via http://alberti.mit.edu/arch/4.207/anneb/thesis/toc.html, chap. 3.

2. "National Public Telecomputing Network Affiliate Systems and Organizing Committees" (National Public Telecomputing Network document, April 27, 1995), available via ftp nptn.org /pub/info.nptn.

3. Cisler, Steve, "Can We Keep Community Networks Running?" *Computer-Mediated Communication Magazine* 2, no. 1 (January 1, 1995), available via http://sunsite.unc.edu/cmc/mag/january.

4. Grundner, Tom, letter to the editor, *Chronicle of Higher Education* (September 7, 1994); in response to "Some Commercial Internet Services Object to Tax-Supported Projects," *Chronicle of Higher Education* (July 27, 1994). See also "Freenets Raise the Ire of Commercial Providers," *Chronicle of Higher Education* (June 29, 1994): A19. All three articles are available via

http://freenet.carleton.ca/freeport/freenet/conference2/issues /menu Directory: Free-Nets are Bad Public Policy.

5. Clarkson University and Partners (principal authors: Ronald Chorba and Wm. Dennis Horn), "North Country Network Project: A Regional Partnership for Rural Access to Technology" (submitted to Telecommunications and Information Infrastructure Assistance Program [TIIAP], National Telecommunications and Information Administration, U.S. Department of Commerce, April 20, 1995).

6. Pearson, Gordon and Brett Dalmage, "National Capital FreeNet 1995 Development Plan," available via telnet freenet.carleton.ca/go funding/Fund Raising Documents.

7. Pearson and Dalmage.

8. Cisler.

9. Telecommons Development Group, "FreeSpace FAQ 2.5," available via http://tdg.uoguelph.ca.

10. "Wellington FreeSpace Mandate," available via http://tdg.uoguelph.ca.

11. Richardson, Don, and Greg Searle, "An Innovative Approach to Rural Community Electronic Networking" (March 1, 1995), available via http://tdg.uoguelph.ca.

12. Miller, Stephen E., "Building the NII from the Bottom Up: A Strategy for Working Through Local Organizations," *Network Observer* (August 1994), available via e-mail rre-request@weber.ucsd.edu, subject: archive send tno-august-1994.

13. Taylor, Richard, "Report on the National Capital FreeNet All Candidates' Meeting for Federal Election 1993," available via telnet freenet.carleton.ca/ About the National Capital FreeNet/Annual Reports/Report on On-Line All Candidates' Meeting.

14. Minnesota Electronic Democracy Project documents, available via http:// free-net.mpls-stpaul.mn.us:8000/govt/e-democracy.

15. Harter, Peter F., "Laws of Electronic Communities and Their Roads: High Noon?" *OnTheInternet* 1, no. 1, available via http://www.nptn.org/cyber. serv/solon/lic/aupessay.html.

16. "The FreeNet Membership Agreement," available via telnet freenet. carleton.ca/Administration/Getting a FreeNet Account.

17. "NCF Policy on Privacy," available via telnet freenet.carleton.ca/All About the National Capital FreeNet/About the National Capital FreeNet/FreeNet Board Business/NCF Policy Statements.

18. Sutherland, Dave, article no. 582, ncf.board-discussion-moderated, available via telnet freenet.carleton.ca/All About the National Capital FreeNet/About the National Capital FreeNet/FreeNet Board Business.

19. "National Capital FreeNet Censorship Policy," available via telnet freenet.carleton.ca/All About the National Capital FreeNet/About the National Capital FreeNet/FreeNet Board Business/NCF Policy Statements.

20. "Guidelines for Debate," available via telnet freenet.carleton.ca/Ontario Provincial Election Project/About the Ontario Provincial Election Project.

### Chapter 11: Reality versus the Communitarian Ideal

1. Cisler, Steve, "Can We Keep Community Networks Running?" *Computer-Mediated Communication Magazine* 2, no. 1 (January 1, 1995), available via http://sunsite.unc.edu/cmc/mag/january.

2. Bromley, Rebekah, "How Subscribers Use the BEV Network," available via http://www.bev.net/about bev page/registration, research, technical reports, publications. Subsequent quotations and results of the survey are from this document.

3. Patrick, Andrew, "Developing a Policy on Telnet Services" (January 27, 1995), available via telnet freenet.carleton.ca/All About the National Capital FreeNet/FreeNet Board Business/Board Meeting Agenda/Selected Documentation from Prior Agendas; Patrick, Andrew, "Internet Service Policy and Equitable Access" (January 27, 1995), available via telnet freenet. carleton.ca/All About the National Capital FreeNet/FreeNet Board Business/Board Meeting Agenda/Selected Documentation from Prior Agendas. Subsequent quotations of Patrick are from "Developing a Policy on Telnet Services."

4. Patrick, "Internet Service Policy and Equitable Access."

5. NCF weekly usage data is available via telnet freenet.carleton.ca/go comnet-sig/netguide.

6. Patrick, Andrew, Alex Black, and Thomas Whalen, "Frequency of Use for Regularly-Used Features on the National Capital FreeNet: Counting 'Go' Commands" (paper presented at the Canadian Community Networks Conference, Ottawa, Ontario, August 15, 1994).

7. Cisler.

8. Clarkson University and Partners (principal authors: Ronald Chorba and Wm. Dennis Horn), "North Country Network Project: A Regional Partnership for Rural Access to Technology" (submitted to Telecommunica-

tions and Information Infrastructure Assistance Program [TIIAP], National Telecommunications and Information Administration, U.S. Department of Commerce, April 20, 1995).

## Chapter 12: "Today's Next Big Something"

1. Etzioni, Amitai, *The Spirit of Community* (New York: Touchstone, 1993), 2, 121.
2. "Study Is Planned on Cable TV Access," *New York Times* (July 14, 1971): 71.
3. Graham, Garth, "A Domain Where Thought Is Free to Roam: The Social Purpose of Community Networks" (paper presented at the CRTC public hearing on information highway convergence, March 29, 1995), available via e-mail aa127@freenet.carleton.ca; CRTC, "The Integration of Cable Television in the Canadian Broadcasting System" (paper prepared for the public hearing held April 26, 1971 [February 26, 1971]): 26, quoted in Gillespie, Gilbert, *Public Access Cable Television in the United States and Canada* (New York: Praeger Publishers, 1975): 2.
4. Center for the Analysis of Public Issues, *Public Issues* supplement no. 1 (July 1971): 1, quoted in Gillespie, 3; Nicholas Johnson, speech to Urban CATV Workshop, Washington D.C., Public Notice, Federal Communications Commission (June 26, 1971): 13, quoted in Gillespie, 8–9; Price, Monroe E., and John Wicklein, *Cable Television: A Guide for Citizen Action* (Philadelphia: Pilgrim Press, 1972), 2; Smith, Ralph Lee, *The Wired Nation, Cable TV: The Electronic Communications Highway* (New York: Harper and Row, 1972), 8.
5. Gillespie, 11.
6. Lappin, Todd, "Deja Vu All Over Again," *Wired* 3.05 (May 1995): 175.
7. Knox, William T., "Cable Television," *Scientific American* 225 (October 1971): 22–29.
8. Rensberger, Boyce, "Cable TV: Two-Way Teaching Aid," *New York Times* (July 2, 1971): 16.
9. Rensberger, 16.
10. Fraser, C. Gerald, "Cable TV Has Block Party," *New York Times* (July 2, 1971): 67.
11. Gent, George, "City Starting Test of Public Cable TV," *New York Times* (July 1, 1971): 95.
12. Gillespie, 6.
13. Fraser, 67.

14. de Sola Pool, Ithiel, "Citizen Feedback in Political Philosophy," in *Talking Back: Citizen Feedback and Cable Technology,* ed. Ithiel de Sola Pool (Cambridge: MIT Press, 1973). The quotation is from page 245.

15. Dansereau, Fernand, *Newsletter Challenge for Change* (NFB) 1 (Winter 1968–1969), quoted in Gillespie, 22.

16. Gwyn, Sandra, "A Report on a Seminar Organized by the Extension Service, Memorial University of Newfoundland, St. John's Newfoundland, March 13–24, 1972," *Film Video-Tape and Social Change* 5, quoted in Gillespie, 27.

17. Bretz, Rudy, *Handbook for Producing Educational and Public-Access Programs for Cable Television* (Englewood Cliffs, N.J.: Educational Technology Publications, 1976), 101–107.

18. Gwyn quoted in Gillespie, 26; Shamberg, Michael, and Raindance Corporation, *Guerrilla Television* (New York: Holt, Rinehart and Winston, 1971), 19, quoted in Gillespie, 37–38.

19. Shamberg quoted in Gillespie, 41.

20. Drew, Jesse, "Media Activism and Radical Democracy," in *Resisting the Virtual Life: The Culture and Politics of Information,* ed. James Brook and Iain A. Boal (San Francisco: City Lights Books, 1995), 77–78.

21. Gillespie, 30.

22. Bretz, 24.

23. Bretz, 23.

24. Price and Wicklein, 18.

25. Minnesota Telecommunications Network document, available via http://www.mtn.org/mtn/index.html.

26. Davitian, Lauren-Glenn, of Chittenden County Television (CCTV), interview by author, July 18, 1995, Burlington, Vermont.

27. Minnesota Telecommunications Network document, available via http://www.mtn.org/mtn/index.html.

28. Davitian.

29. Kalba, Kas, "The Electronic Community: A New Environment for Television Viewers and Critics," in *Television as a Social Force: New Approaches to TV Criticism,* ed. Douglass Cater, Aspen Series on Communications and Society (New York: Praeger Publishers, 1975), 159.

## Chapter 13: Fight the Good Fight

1. Old North End Community/Tech Center newsletter (July 1995), information available via e-mail ctc@cctv.uvm.org.
2. Old North End Community/Tech Center newsletter.
3. Bradshaw, Chris, "Exploring the Concept of the Neighbourhood Computer Network ('Neigh-Net')" (December 4, 1994), available via telnet freenet.carleton.ca/go comnetsig/5. making connections in the community/12. NCF birthday proposal for neighborhood nets. Subsequent quotations of Bradshaw are from this document.
4. Schwartz, Ed, "The Neighborhood Agenda" (September 1994), available via gopher.civic.net 2400.
5. Morino Institute, "The Promise and Challenge of a New Communications Age" (May 15, 1995), available via http://www.morino.org. All quotations in the following discussion of the Morino Institute's suggestions are from this document.
6. Stoll, Cliff, *Silicon Snake Oil* (New York: Doubleday, 1995).

# Index